The Age of
REASON

1600 TO 1750

Published by The Reader's Digest Association Limited
London • New York • Sydney • Montreal

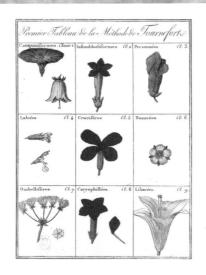

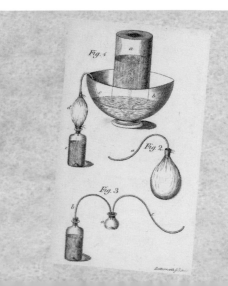

Contents

Introduction   4

THE STORY OF INVENTIONS   18
Newspapers   20
Natural rubber   24
Horse-drawn cabs   26
THE COPERNICAN SYSTEM   27
Underwater exploration   32
GALILEO   36
Circulation of the blood   40
The flintlock   44
The vernier calliper   47
The umbrella   47
Pascal's calculator   48
RENÉ DESCARTES   54
The barometer   58
The grenade   60
The post box   61
The vacuum pump   61
Banknotes   62
The microscope   67
ANTON VAN LEEUWENHOEK   72
Fire insurance   76
The spirit level   78
Street lighting   78
The Roberval balance   79
CHRISTIAAN HUYGENS   80

The reflecting telescope   82
Binoculars   88
Lead glass   89
Ice cream   89
Botanical classification   90
Universal gravitation   94
ISAAC NEWTON   97
Champagne   101
ADVANCES IN ANATOMY   102
The steam engine   108
The seed drill   114
EDMUND HALLEY   116
The clarinet   119
The pianoforte   120
ACADEMIES AND SALONS   124
The tuning fork   129
Statistics   130
Eau de Cologne   132
Central heating   132
4-colour printing   133
The sextant   134
The first android   136
Sparkling water   137
The capacitor   137
Sugar beet   138
COSMOPOLITAN AMSTERDAM   140

CHRONOLOGY   144
Index   154
Acknowledgments   159

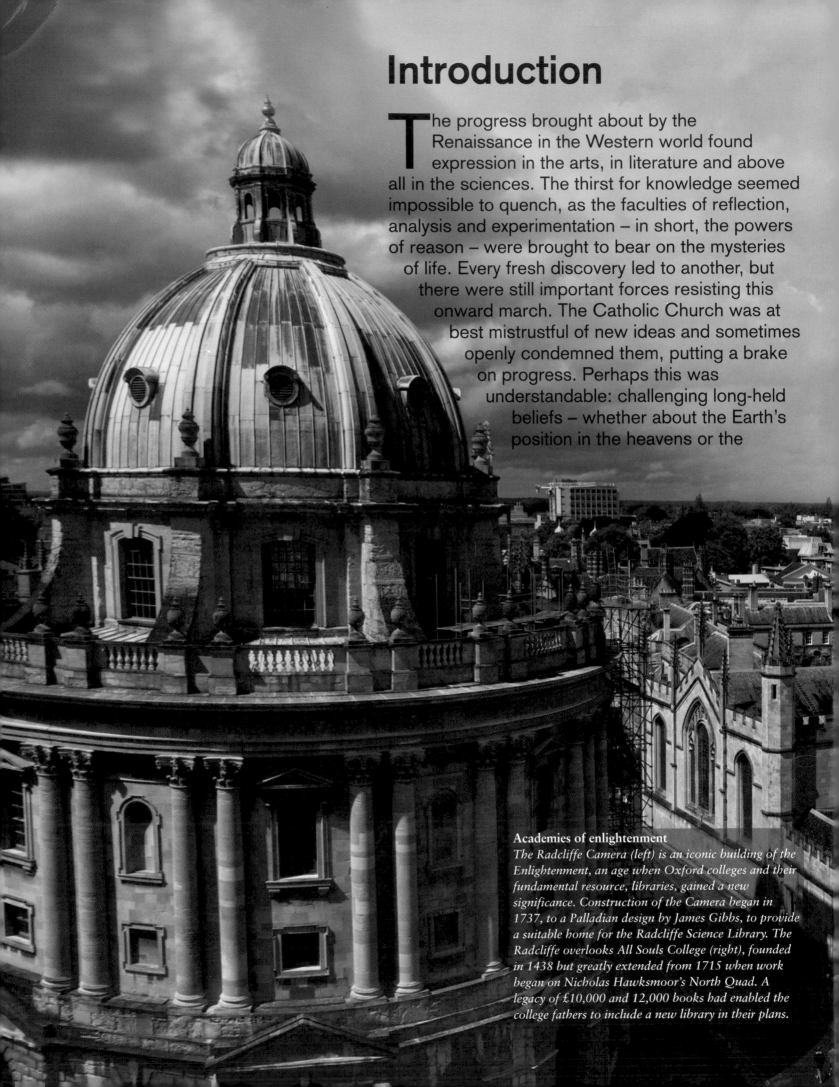

Introduction

The progress brought about by the Renaissance in the Western world found expression in the arts, in literature and above all in the sciences. The thirst for knowledge seemed impossible to quench, as the faculties of reflection, analysis and experimentation – in short, the powers of reason – were brought to bear on the mysteries of life. Every fresh discovery led to another, but there were still important forces resisting this onward march. The Catholic Church was at best mistrustful of new ideas and sometimes openly condemned them, putting a brake on progress. Perhaps this was understandable: challenging long-held beliefs – whether about the Earth's position in the heavens or the

Academies of enlightenment
The Radcliffe Camera (left) is an iconic building of the Enlightenment, an age when Oxford colleges and their fundamental resource, libraries, gained a new significance. Construction of the Camera began in 1737, to a Palladian design by James Gibbs, to provide a suitable home for the Radcliffe Science Library. The Radcliffe overlooks All Souls College (right), founded in 1438 but greatly extended from 1715 when work began on Nicholas Hawksmoor's North Quad. A legacy of £10,000 and 12,000 books had enabled the college fathers to include a new library in their plans.

structure of the human body – raised questions about the fundamental nature of humankind, its relationship with God and our place in the Universe. But in the long run change would come regardless, even if some bold pioneers paid dearly for their audacity: Giordano Bruno died at the stake for his ideas in 1600.

As the spirit of scientific enquiry blossomed, first of all in the Protestant lands of northern Europe, scholars gained a new status: once restricted to theological studies, wrapped in mystery and shielded from public scrutiny, they emerged to claim a new place in society. By order of the Inquisition, Galileo was under house arrest at the time of his death in 1642. Yet Isaac Newton, born that same year, was able to enjoy a brilliant career as a public figure. Descartes published his ground-breaking philosophical works in the Netherlands, rather than his native France, but his ideas were soon familiar across Europe.

The Scientific Revolution of the 17th and early 18th centuries was universal in its ambitions. Its tool was mathematics, believed to be the quintessential expression of the rational mind, which emerged from the shadow of philosophy to become a guiding light. If everything could and should be expressed in numbers, then all could be measured, too. The invention of the telescope and microscope, vernier calliper and sextant, owed little to chance and almost everything to careful calculation.   *The editors*

▼ *Nieuwe Tijdinghe* ('New Tidings') was an illustrated weekly periodical, with numbered pages. First published in Amsterdam early in the 17th century, it was the ancestor of modern newspapers.

▼ In 1640 a French coach-maker named Nicolas Sauvage opened an office in Paris from which people could rent carriages by the hour or by the day. The horse-drawn cab had put in its first appearance.

► In 1620 the Dutch inventor Cornelis Drebbel built the first submersible craft out of wood and leather. Propelled by 12 oars, it travelled for 4 miles below the surface of the River Thames in London.

Classical writers used to claim that man is the measure of all things. Pious medieval scholars added that God alone should be our reference point. But by the year 1600, a change was taking place: a new age was dawning in which the power of reason

◄ The Polish astronomer Copernicus set the scene for the Age of Science with his ground-breaking *De Revolutionibus*, published in 1543, the year of his death. In it he proposed a heliocentric model of the solar system, contradicting the view of Ptolemy and other Classical astronomers that the Earth was the centre of the universe.

▼ William Harvey explains to King Charles I the principles of the circulation of the blood and the heart's role as a pump. Harvey was the royal physician as well as a scientific pioneer.

► Probably invented by the Greek astronomer Eratosthenes in the 3rd century BC, the armillary sphere continued in use into the 17th century. It showed the principal divisions of the skies and movements of the heavenly bodies.

would be the guiding principle. Or so, at least, proponents of the 'Scientific Revolution' believed. In practice, it turned out that there were still many battles to be fought to break free from old ways. William Harvey was among those who found this out the

▼ As a professor at the University of Padua, Galileo (1564–1642) proved to be an exceptional teacher. A critic of medieval scholasticism, he improvised his lessons and staged experiments as teaching aids.

► William Harvey included this illustration in his ground-breaking work on the circulation of the blood, showing how ligatures could be used to make veins stand out.

◄ The telescopes built by the Dutch pioneers magnified objects three times. Galileo borrowed the idea to develop astronomical instruments that provided 30x magnification. These examples are from 1609–10.

TVBVM·OPTICVM·VIDES·GALILAEII·INVENTVM.ET·OPVS,QVO·SOLIS·MACVLAS, ET·EXTIMOS·LVNAE·MONTES·ET·IOVIS·SATELLITES,ET·NOVAM·QVASI RERVM·VNIVERSITATE·PRIMVS·DISPEXITA.MDCIX.

hard way: having proved that the heart pumps blood around the body, he then struggled to get his ideas accepted. Meanwhile, many doctors continued the outdated practice of bleeding their patients, even though Harvey's discoveries had rendered such

► To demonstrate the effect of atmospheric pressure, the French philosopher-scientist Blaise Pascal had the idea of placing a barometer on the 1,464m summit of the Puy-de-Dôme in the Massif Central.

▼ In 1642 Pascal invented this calculating machine, whose inner workings featured an ingenious arrangement of toothed wheels.

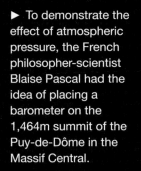

▲ Between 1608 and 1615 a French gunmaker, Marin le Bourgeoys, perfected a firing mechanism using flint to strike sparks. The result was the flintlock musket, which was both cheaper and easier to handle than earlier guns. Corps of fusiliers were formed to use them.

treatment obsolete. Shaken by the impact of humanism and the Reformation, the guardians of Catholic orthodoxy took refuge in tradition. Calling everything into question is, after all, never comfortable. The Age of Reason would not come into being

▲ The French philosopher René Descartes at the court of Queen Christina of Sweden (top, third from right). Descartes posed fundamental questions of existence in his *Principia Philosophiae* ('Principles of Philosophy'), published in 1644 (above).

▶ The first European banknotes were issued in Sweden in 1661. This one-guinea note, distributed by the Bank of Scotland in 1777, was printed in colour to make it harder to counterfeit.

without some birth-pangs. Even so, the scientific pioneers had one big advantage in their favour, which was that their discoveries quickly brought inventions and technological advances in their wake to the benefit of everyone. As René Descartes, the French

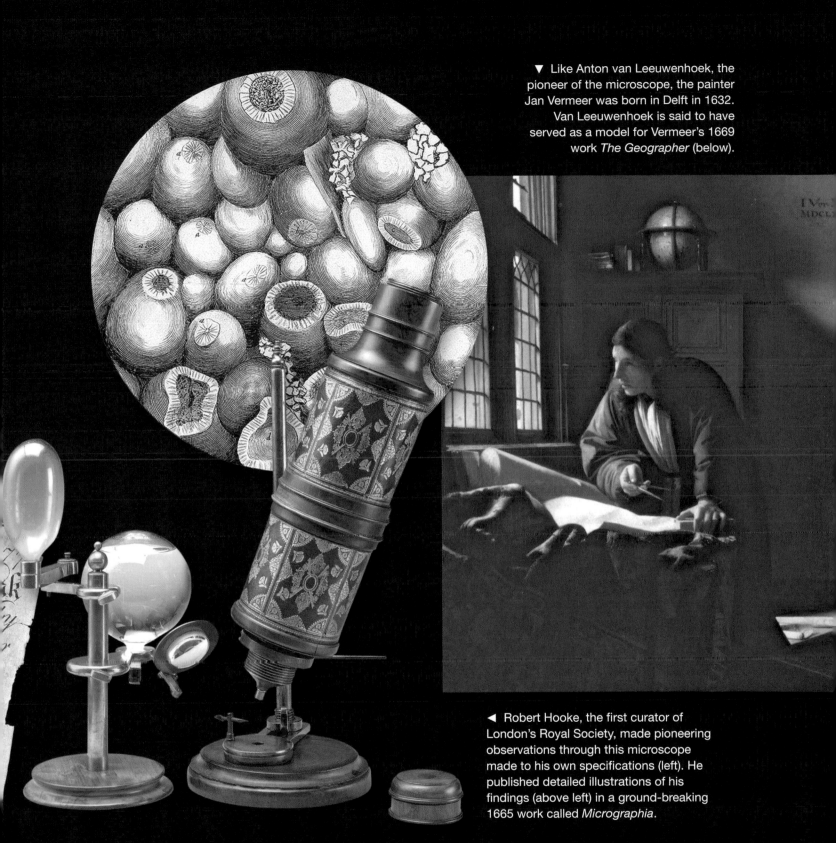

▼ Like Anton van Leeuwenhoek, the pioneer of the microscope, the painter Jan Vermeer was born in Delft in 1632. Van Leeuwenhoek is said to have served as a model for Vermeer's 1669 work *The Geographer* (below).

◀ Robert Hooke, the first curator of London's Royal Society, made pioneering observations through this microscope made to his own specifications (left). He published detailed illustrations of his findings (above left) in a ground-breaking 1665 work called *Micrographia*.

philosopher of science, pointed out, the exercise of reason would lose all purpose without the knowledge gained from direct experience, and this was a time when scientists were not just mathematicians, physicists and philosophers, but also in many

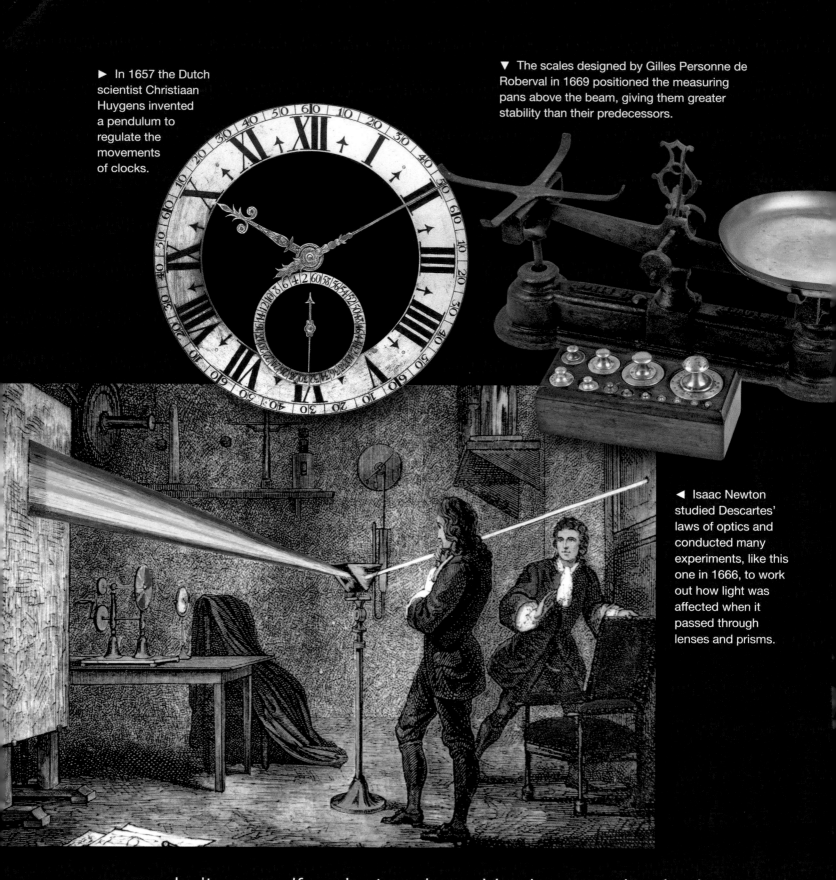

► In 1657 the Dutch scientist Christiaan Huygens invented a pendulum to regulate the movements of clocks.

▼ The scales designed by Gilles Personne de Roberval in 1669 positioned the measuring pans above the beam, giving them greater stability than their predecessors.

◄ Isaac Newton studied Descartes' laws of optics and conducted many experiments, like this one in 1666, to work out how light was affected when it passed through lenses and prisms.

cases do-it-yourself gadget-makers. Van Leeuwenhoek pioneered microscopy, discovering spermatozoa and the red cells in blood, because he enjoyed messing around with magnifying glasses. Newton not only propounded the law of universal gravitation but

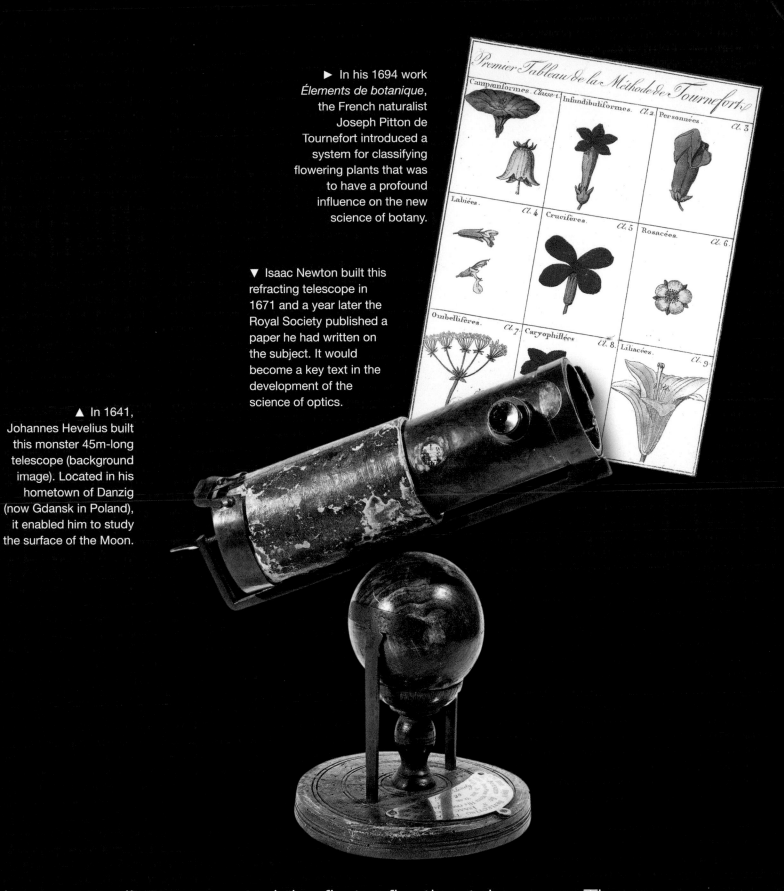

▶ In his 1694 work *Élements de botanique*, the French naturalist Joseph Pitton de Tournefort introduced a system for classifying flowering plants that was to have a profound influence on the new science of botany.

▼ Isaac Newton built this refracting telescope in 1671 and a year later the Royal Society published a paper he had written on the subject. It would become a key text in the development of the science of optics.

▲ In 1641, Johannes Hevelius built this monster 45m-long telescope (background image). Located in his hometown of Danzig (now Gdansk in Poland), it enabled him to study the surface of the Moon.

also personally constructed the first reflecting telescope. The teenage Pascal developed his calculating machine while helping his father with the accounts. In this golden age of discovery there was hardly a step forward in pure science that did not soon turn

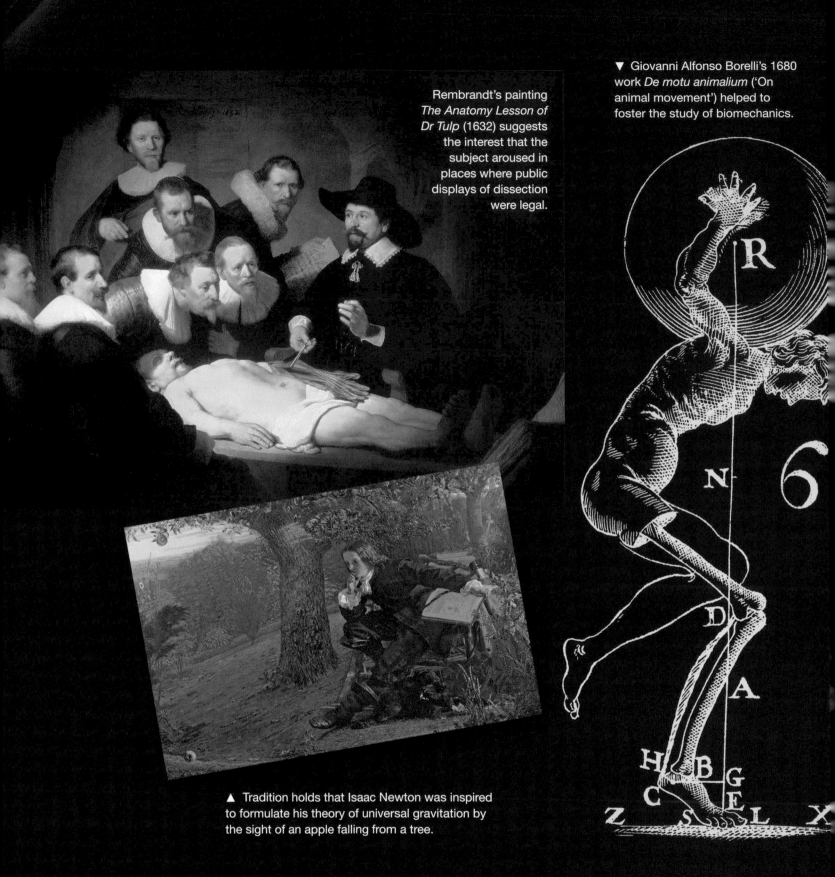

Rembrandt's painting *The Anatomy Lesson of Dr Tulp* (1632) suggests the interest that the subject aroused in places where public displays of dissection were legal.

▼ Giovanni Alfonso Borelli's 1680 work *De motu animalium* ('On animal movement') helped to foster the study of biomechanics.

▲ Tradition holds that Isaac Newton was inspired to formulate his theory of universal gravitation by the sight of an apple falling from a tree.

up some practical application. The best example of all, maybe, is the steam engine. Invented by Denis Papin as an offshoot of his research into vacuums, it owed its global uptake to the many different uses it could be put to. In time it became more than just a

▼ Denis Papin was the first person to demonstrate the principle of the steam engine, in 1687.

▼ In 1698 Thomas Savery patented a steam pump based on the principles that Papin had demonstrated.

▼ In the final years of the 17th century Edmund Halley became the first person to measure and record magnetic declinations, publishing the results in a 1701 world map.

machine, it was the key to the Industrial Revolution. So it was that the 17th and early 18th centuries marked the start of one of history's most fruitful ages, when scientific and technological progress went hand in hand. Between them, physics and maths

15

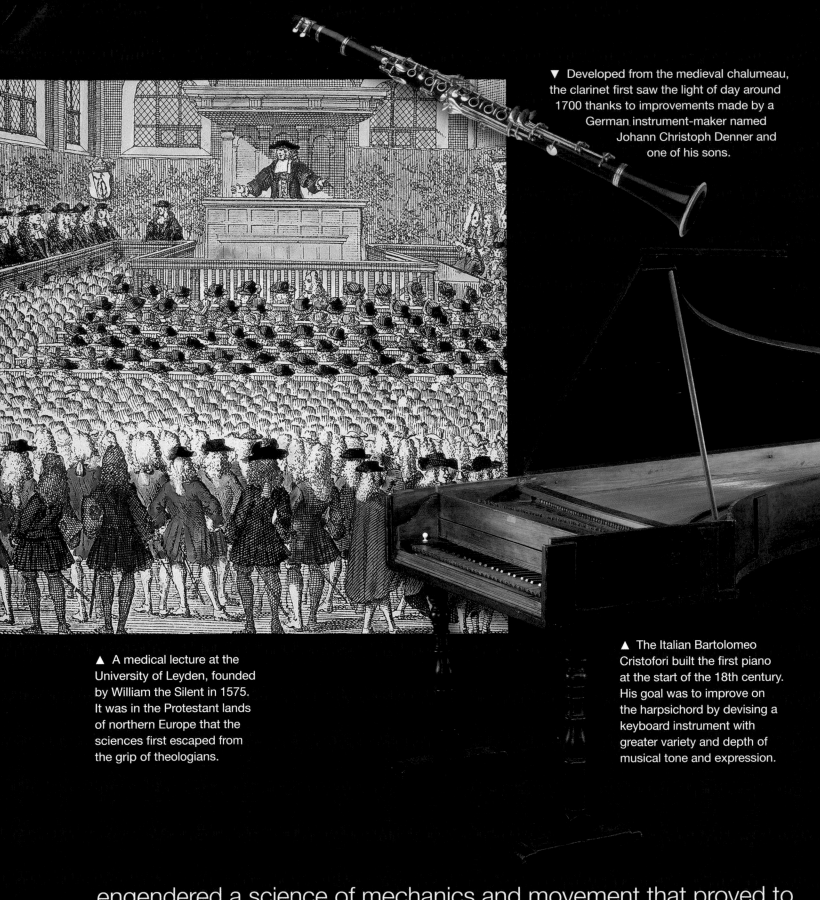

▼ Developed from the medieval chalumeau, the clarinet first saw the light of day around 1700 thanks to improvements made by a German instrument-maker named Johann Christoph Denner and one of his sons.

▲ A medical lecture at the University of Leyden, founded by William the Silent in 1575. It was in the Protestant lands of northern Europe that the sciences first escaped from the grip of theologians.

▲ The Italian Bartolomeo Cristofori built the first piano at the start of the 18th century. His goal was to improve on the harpsichord by devising a keyboard instrument with greater variety and depth of musical tone and expression.

engendered a science of mechanics and movement that proved to be the engine of progress and improved standards of living. For a time science seemed so omnipotent that it threatened to dominate all human activity. Fortunately, even when reason seemed to be in

▼ In 1745 Andreas Marggraf found a way to transform beet into sugar. The process did not become a commercial venture until Napoleon's Continental System cut much of Europe off from Caribbean cane sugar.

▼ In 1731 John Hadley devised the octant to measure latitude at sea. It served as the model for the later sextant.

► From its origins as a small herring port, Amsterdam grew into an economic and cultural powerhouse in the 17th century.

total control, human nature continued to be multi-faceted. The age of science and reason was also the era when ice-cream, champagne, the clarinet and the piano were invented. Even at their most rational, people continued to have a taste for life's pleasures.

THE STORY OF INVENTIONS

The world was on the move as the 17th century began. Novel ideas and fresh thinking were in the air. Newspapers and journals were taking their place alongside scholarly works as vehicles for disseminating the new knowledge. In princely courts and academies, people were eagerly exchanging theories. Inventions quickly followed as ingenious individuals found ways to apply the results of their researches. Technology opened up new perspectives previously only glimpsed by artists and poets, but now brought into the world by men of science.

Information takes power

Hand-written news-sheets date back thousands of years, but the 17th century saw the birth of the printed newspaper, inaugurating the era of modern mass communications. The first journals appeared in Holland, then other European nations quickly followed suit, encouraging the development of a new socio-political phenomenon – public opinion.

Pioneer press
The earliest surviving copy of Nieuwe Tijdinghe *(below) dates from 1621.*

In May 1605 a printer named Abraham Verhoeven, living in Antwerp, then in the southern Netherlands, was granted the privilege to 'print and engrave on wood or metal and to offer for sale … all the most recent news'. A few years later a weekly publication entitled *Nieuwe Tijdinghe* ('New Tidings') first saw the light of day. It took the form of an eight-page illustrated journal in octavo format (about half the size of an A4 sheet of paper), and its regular publication and the variety of the subjects it covered earned it the right to be considered the ancestor of the modern newspaper. It had precursors of its own, notably commercial gazettes aimed at the merchant community in Antwerp and Venice, but these were intended for a specialised rather than a general readership. In 1629 *Nieuwe Tijdinghe* was replaced by another, very similar, periodical called *Wekelijke Tijdinghe* ('Weekly Tidings'), which was published from 1629 to 1632.

Ancestors of the newspaper

The concept of news-sheets was already well established long before Verhoeven first set to work. Way back in Roman times some 10,000 copies of a hand-written leaflet called *Acta Diurna* ('Daily Events') were distributed around the empire from the mid 1st century BC. In China gazettes called *pao* circulated in court circles from AD 713, surviving in an unchanged format to as late as 1911. From the early 18th century Germans could read *zeitungen*, concerned above all with military matters. And in medieval Venice freelance reporters produced *fogli a mano* ('manuscript sheets') for merchants and bankers.

By the 15th century, following the onset of the Renaissance and the invention of printing, a variety of publications covering current affairs began to appear at irregular intervals. Some were broadsheets concerned with topical issues, others were pamphlets or lampoons whose partisanship presaged the development of modern editorial and opinion pages. The spread of literacy across the Continent provided such publications with a growing readership, while the development of postal services facilitated distribution. News reporting – contemporary, eye-witness accounts of events – began in 1513 with the report by Richard Faques of the defeat of James IV of Scotland by the English army at the Battle of Flodden. It was published in a pamphlet entitled *The True Encounter*.

Meanwhile, almanacs set a precedent as the first periodical publications. Intended for a

SHARING THE NEWS

The high price of early newspapers – *Nieuwe Tijdinghe* cost 2 sous, for example, as much as a large loaf of bread – meant that most people read them in cafés or clubs, which attracted a growing clientele as a result.

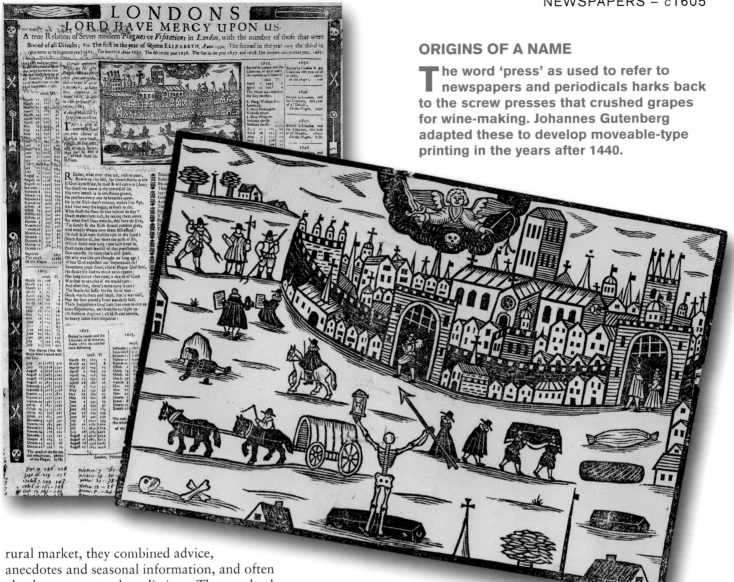

ORIGINS OF A NAME

The word 'press' as used to refer to newspapers and periodicals harks back to the screw presses that crushed grapes for wine-making. Johannes Gutenberg adapted these to develop moveable-type printing in the years after 1440.

rural market, they combined advice, anecdotes and seasonal information, and often also horoscopes and predictions. They evolved from astronomical almanacs thanks to the work of scholars such as Roger Bacon, whose calendar appeared in 1292.

Rapid expansion

In the years after 1600 the next logical step was to produce news-sheets to a set schedule. The newspaper was born. The concept soon spread beyond Antwerp to Frankfurt, Amsterdam, Florence, Berlin, Madrid, Paris and London. Would-be journalists in London risked drawing down the ire of the Court of Star Chamber, with the result that the first English-language newspaper was in fact printed and published in Amsterdam, in 1620. In 1632 'news books' were banned completely and the birth of anything resembling a free press had to await the downfall of Charles I. In colonial America a journal called *Public Occurrences* appeared in Boston in 1690; it soon closed but the *Boston Newsletter*, established in 1704, continued to publish until the British withdrawal 72 years later.

These early gazettes at first limited their scope and ambitions to recording official pronouncements, but they soon also started running stories and commentaries on other aspects of life. In the early days they were constantly hampered by the threat of official interference. The state sought to exercise control by forcing them to obtain licenses to publish, as well as through direct censorship, and it also tried to exploit the

Illustrating the news
A news-sheet from 1665, the time of the Plague in London, featured woodblock print illustrations. This one shows burying the dead.

EYES OF THE SPECTATOR

One of the most influential of the early English newspapers was the *Spectator*, which was published daily for 555 issues in 1711 and 1712. It was revived two years later, when a further 80 issues appeared. The *Spectator* addressed social and cultural issues rather in the manner of modern lifestyle magazines, but did so through the eyes of a cast of fictional personalities, character types who were members of an imaginary Spectator Club. The best-remembered of them is Sir Roger de Coverley, a lovable, old-fashioned country gentleman. The review was the brainchild of Joseph Addison and Richard Steele, friends from schooldays; other contributors included the poet Alexander Pope.

THE FATHER OF FRENCH JOURNALISM

In 1629 Théophraste Renaudot opened a pioneering employment agency in Paris to bring together job-seekers and would-be employers. The agency also advertised tradesmen's services, eventually building up a directory of half a million addresses. With the support of France's chief minister, Cardinal Richelieu, Renaudot went on to found France's first newspaper, *La Gazette*, in 1631, taking the name from the *gazetta,* an Italian coin that happened to be the price of a weekly news-sheet in Venice. Richelieu used the publication to promote his foreign policy, particularly his controversial decision to form alliances with Protestant German states in the Thirty Years' War. The paper, which won praise from such leading figures as the playwright Molière, lasted almost three centuries until 1915. A decade later French journalists named one of France's premier literary awards the Prix Renaudot, after their early patron.

Dual purpose platform
Renaudot's employment bureau also served as a forum for debate, hosting weekly conferences on a variety of topical subjects.

fledgling press for its own purposes. Théophraste Renaudot, considered the father of journalism in France (see box above), summed up the limits of the new craft: 'In one matter I will not yield to any man, and that is in the pursuit of truth; yet, even so, I cannot stand surety for it.' Meanwhile, pamphlets and other one-off publications that maintained a freer tone continued to circulate, often clandestinely.

Liberty print
The French Revolution declared liberty of the press to be a fundamental right, citing it in Article 11 of the Declaration of the Rights of Man. This satirical engraving (left) purports to show what happens when, in the words of the Declaration, *'every citizen can speak, write and print in freedom'.*

Press censorship could not prevent the appearance of an ever-growing number of titles over the course of the 17th and 18th centuries. Among them were specialised publications devoted to literature, science and social gossip. Daily newspapers soon followed. One of the first to come off the presses was the *Leipziger Zeitung*, which was launched in 1660 only to fold soon after. The true pioneer was Britain's *Daily Courant*, which was published in London's Fleet Street from 1702 until 1735, when it merged with the *Daily Gazetteer*.

The fight for a free press

From the start, journalism struggled to escape the grip of the powers-that-be. The task was accomplished at different times in different countries. In relatively liberal societies such as Britain and the Low Countries or, slightly later, the youthful USA, newspapers acquired a degree of freedom as early as the 18th century. *The Times*, the oldest surviving paper in England, was originally published in 1785 as the *Daily Universal Register*. It acquired its new name in 1788, though it was also known as 'The Thunderer' for its outspoken attitude. Meanwhile, the press was rigidly censored in more authoritarian states such as Prussia and France. On-going state interference helped to explain why Enlightenment philosophers such as Jean-Jacques Rousseau held newspapers in such disdain – 'ephemeral works' fit only 'to provide women and fools with trivialities'.

The principle of a free press only became firmly established in the Western world with the spread of democracy from the 19th century on. Even then, all was not necessarily well. Having won their freedom, newpapers became objects of mass consumption, partly thanks to technological advances that cut production costs. Circulations mushroomed, and before long the popular press was rubbing shoulders with publications offering more intellectually demanding fare. Journalists became respected professionals, sometimes playing an important part in politics. The years between 1880 and 1930 were a golden age for newspapers. Thereafter, first radio, then television ate into the press's monopoly on news reporting. Now, four centuries after the birth of the newspaper, the fear is that electronic media may be sounding the death-knell of daily and weekly papers.

THE FOURTH ESTATE

The term 'fourth estate', as applied to newspapers, was coined in 1787 by the Irish-born statesman and political theorist Edmund Burke. The British press had been covering parliamentary debates since 1771, and its influence had grown to the point where, in Burke's view, it rivalled that of the 'three Estates in Parliament' – the Church, Lords and Commons. Looking up at the reporters' gallery in the House he declared, 'Yonder, there sat a Fourth Estate more important far than they all'.

Front-page scoop
The Washington Star-News *of August 9, 1974, reports the imminent resignation of President Nixon, which was announced later that evening on television.*

Delivery chain
A team of paperboys clutching newspaper bundles outside a distribution centre in Boston, at 5 o'clock one October morning, 1909. Their working day had already begun.

Natural rubber c1615 AD

Europeans first came across rubber as a result of the Spanish conquest of South America, where indigenous Amazonian peoples knew rubber trees as 'the wood that weeps'. Columbus brought back samples at the end of the 15th century, and the substance also found its way into the holds of vessels bringing conquistadors back from central America. Even so, the viscous white fluid remained unknown to most Europeans at the time.

Aztec ingenuity

In 1530 the Italian geographer Pietro Martire d'Anghiera wrote of the rubber tree in his book *De Orbo Novo* ('From the New World'), saying that: 'From one of these trees a milky sap exudes. Left standing it thickens to a kind of pitchy resin.' A Spanish historian, Antonio de Herrera Tordesillas, also wrote a description of the sap of the India rubber tree and was possibly the man who introduced the substance to the Old World. Tordesillas described how the Aztecs used the sap for many purposes: in particular, to make shoes, jackets and capes that provided protection from bad weather. There were also some less expected uses, such as to heal coughs and keep babies warm. And rubber was used to make balls for the Mesoamerican ballgame, a ritual sport with similarities to modern Basque pelota.

Rich harvest

Another century passed before rubber was mentioned again, this time by the French naturalist Charles-Marie de la Condamine. Travelling through South America in the 1730s, Condamine journeyed down the River Amazon, where he studied the Pará rubber tree, *Hevea brasiliensis*. Later, he met a court-appointed engineer named Francois Fresneau in French Guiana and passed on his enthusiasm for the species.

On his return to France, Fresneau published a scientific paper in 1649 describing the tree's characteristics and the way in which native peoples obtained its sap. He waxed lyrical in praise of rubber's potential uses, claiming among other things that by combining it with

Rubber and ritual
The Aztecs knew of rubber before the arrival of Europeans in Mexico. In this illustration (above) from the Codex Magliabechiano – *an Aztec religious document made in the mid 16th century – Mayahuel, the Aztec goddess of fertility, holds a sceptre decorated with fronds from a Pará rubber tree.*

THE LATEX TREE

Bearing a vague resemblance to the plane tree, the Pará rubber tree, *Hevea brasiliensis*, is the world's main source of the latex used to make natural rubber. Other plants secrete latex – including the guayule, a shrub found in Mexico, and Russian dandelion (*Taraxacum koksaghyz*) – but it is less pure than that found in the Pará rubber tree, containing resins that have to be removed before the rubber can be put to use.

AN EMPIRE OF RUBBER

As its name suggests, *Hevea brasiliensis* is native to South America, yet its cultivation became one of the main resources of the British Empire. The man responsible was an explorer, Henry Wickham, who in 1876 brought 70,000 seeds from Brazil to the Royal Botanic Gardens at Kew. From there, seedlings were dispatched to British colonies in South and Southeast Asia, which soon came to eclipse South America's production.

fabrics manufacturers could 'make diving suits, flasks for wine, bags for soldiers' rations etc., without any fear that the material would impart an unpleasant odour'. But in practice, Brasilian latex proved insufficiently hardy to stand up to the lengthy trans-Atlantic sea journey, reaching Europe in solidified form.

A productive accident

In 1768 two French chemists, Jean-Thomas Hérissant and Pierre-Joseph Macquer, discovered that hardened latex regained its elasticity if it was soaked in ether or in turpentine. The next advance came from a British manufacturer named Samuel Peal, who in 1791 patented a method of waterproofing cloth by treating it with a solution of latex dipped in turpentine. Then, in 1823, the Scottish chemist Charles MacIntosh found the ideal solvent for natural rubber: naphtha, a by-product of coal tar. That same year MacIntosh opened the first factory to make waterproof fabrics. The mac had been born.

Yet problems still remained. For all its adaptability, natural rubber remained sensitive to heat, which softened it, and to cold, which turned it rigid. For five years the American businessman and inventor Charles Goodyear sought a way to stabilise the material without losing its pliancy. One day he accidentally dropped a mixture of gum and sulphur onto a hot stove, and found to his surprise that it remained firm yet retained its elasticity. From this chance discovery in 1839 the process known as vulcanisation came into general use, named after the Roman god of fire. In the wake of Goodyear's discovery, patented in 1844, the rubber industry expanded rapidly, encouraged by a growing demand for tyres, first for bicycles and then for cars. Soon rubber plantations were springing up in every tropical clime. Progress was only checked in the 20th century by the invention of synthetic rubber, which put a cap on the demand for the natural product.

Cultivation in Malaysia
Malaysia is the world's third-largest producer nation. Rubber trees like this one (right) are tapped every second day and the sap is collected in buckets.

Rubber products
Today, Pirelli are best known for their tyres and celebrated calendar, but the image below comes from a Pirelli advertisement in the 1920s.

ERASING ERRORS

For centuries artists had used wax tablets or lumps of bread to erase false strokes. In 1770 the English chemist Joseph Priestley noticed that latex gum was 'excellently adapted to the purpose of wiping from paper the mark of black lead pencil'. He called the substance 'rubber' for its utility in rubbing out mistakes, and the name stuck. It was through the eraser, to give the rubber its posher name, that Europeans first became generally aware of rubber.

Horse-drawn cabs c1617

At the start of the 17th century the normal way of travelling around cities – at least for those who could afford it – was to be carried in sedan chairs. Then, in about 1617, an Amiens coach-maker by the name of Nicolas Sauvage had the idea of providing cabs for hire, the ancestors of the modern taxi. In 1640 he opened a carriage rental business, offering vehicles by the hour or by the day. His offices were located in a district of Paris that formed part of the estate of the former Hotel Saint-Fiacre, and the cabs themselves subsequently became known as fiacres.

Sauvage's fleet comprised about 20 vehicles, but he neglected to obtain an exclusive permit (the equivalent of a patent), and so soon found himself facing competition. It remains unclear whether these first cabs waited for custom at central locations in the city or if clients had to go to the depot to find one. What is certain is that demand was such that a guild of 'carriage-renters' was established in 1718.

From cabs to taxis

Soon entrepreneurs seeking to hire out cabs had to pay a tax for the privilege. The police also imposed regulations on the new business, so that from 1699, for example, it became illegal

Cabs and cabbies
By the time this illustration (above) was made in the 18th century, city-dwellers were already familiar with traffic jams, as the 17th-century illustration of the Pont Neuf in Paris (below) suggests.

to feed horses in the street or to employ drunks or vagabonds as coachmen. From 1703, every Paris cab had to display a readable number – an early form of vehicle registration.

From the first decade of the 20th century horse-drawn cabs found themselves increasingly replaced by motor taxis. In London the change was relatively abrupt; the first petrol-powered vehicles made their appearance in 1903, and by 1910 they already outnumbered the old hansom cabs by some 6,300 to 5,000.

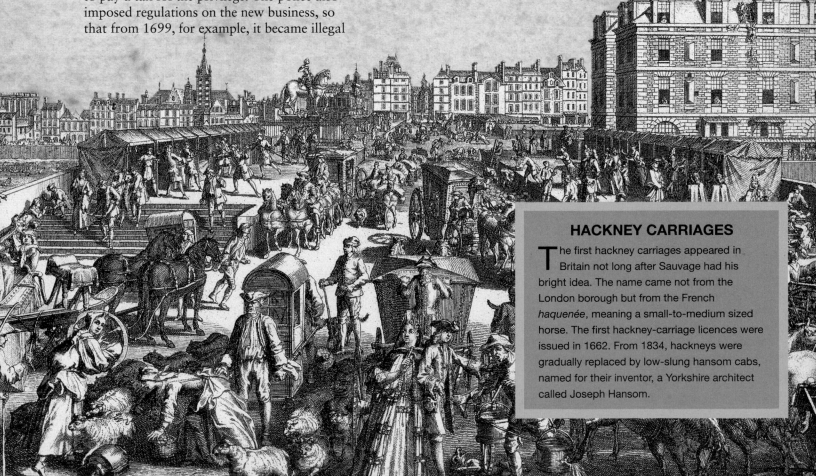

HACKNEY CARRIAGES

The first hackney carriages appeared in Britain not long after Sauvage had his bright idea. The name came not from the London borough but from the French *haquenée*, meaning a small-to-medium sized horse. The first hackney-carriage licences were issued in 1662. From 1834, hackneys were gradually replaced by low-slung hansom cabs, named for their inventor, a Yorkshire architect called Joseph Hansom.

Goodbye to an Earth-centred universe

The Italian scholar Giordano Bruno drew on the work of the Polish astronomer Copernicus to argue that the universe was infinite. The Inquisition disagreed and sentenced him to be burned at the stake. Yet despite the dangers, 17th-century scientists continued to put forward a vision of the cosmos that made the Earth and the heavens all part of a single system. In effect, they prepared the way for the emergence of modern science.

Giordano Bruno went to his death in the Campo dei Fiori, a piazza in Rome, on February 17, 1600. He had been condemned to the flames four weeks earlier on the orders of Pope Clement VIII. A philosopher and theologian as well as a man of science, Bruno had taken issue with the Aristotelian view of the cosmos, which proposed a hierarchical universe with the Earth at its centre and the heavenly bodies on its periphery. This view had been adopted as Church dogma, and merely to question it was seen as a form of blasphemy. Once accused, he had refused to recant his views, telling his judges, 'You who condemn me probably feel more fear than I do'.

Bruno's heresy

In Aristotle's schema, humankind occupied centre stage in the celestial drama. But ever since Copernicus's ideas had been published in 1543, the Sun had increasingly become the focus of scientific attention. The outer heavens were the domain of stars, traditionally viewed as incandescent dots of pure quintessence. Bruno denied that there was any such thing as quintessence; instead, he held that each star was another sun harbouring inhabited worlds like our own – a concept that he labelled 'multiple worlds'. Copernicus's own views seem timid in comparison, doing little more than inverting the previous consensus by placing the Sun in the Earth's old place at the centre of the universe. In Bruno's view, there was no centre and no celestial hierarchy. Instead, there was a multitude of solar systems.

Bruno was ambitious – too much so for his own good. He was also ahead of his time. The people who condemned him no doubt hoped that his cosmological speculations would be consumed along with his mortal remains that February day in 1600. In fact it was the Aristotelian view that was on its way out.

The face of change
Nicholas Copernicus (left) gazes out from a portrait by an unknown artist painted in 1575, more than three decades after the astronomer's death. The 18th-century illustration below depicts the Copernican concept of the universe, showing the Sun firmly ensconced at the centre.

The flame of the new science would burn brighter, lighting up the new century that was just dawning.

The legacy of Copernicus

The initial spark had come from Poland. Between 1512 and 1543 Nicolas Copernicus, a Catholic cleric, developed a heliocentric model of the universe which broke with the Earth-centred tradition of the ancient world as elaborated by the late-Classical cosmographer Ptolemy in the 2nd century AD. Copernicus set out his ideas in a book entitled *De revolutionibus orbium coelestium* ('On the Revolutions of the Celestial Spheres'), published in the year of his death.

To the last Copernicus remained cautious about upsetting Church dogma and the received opinions of his day. Taken out of their historical context, the discoveries that he described may seem insufficient today to explain the extraordinary effect they produced on scholars at the time. But a spark will catch light if there is inflammable material to hand. The work of the reticent cleric was revolutionary in its effects thanks to what had gone before and what was to come after.

For all his importance as a pioneer, Copernicus did not make a clean break with earlier views of the workings of the universe. Certainly, his system denied the Earth the central place it had previously been accorded, making it merely one planet among many. It also did away with the dichotomy between so-called sublunary physics, concerned only with what happened on Earth, and the

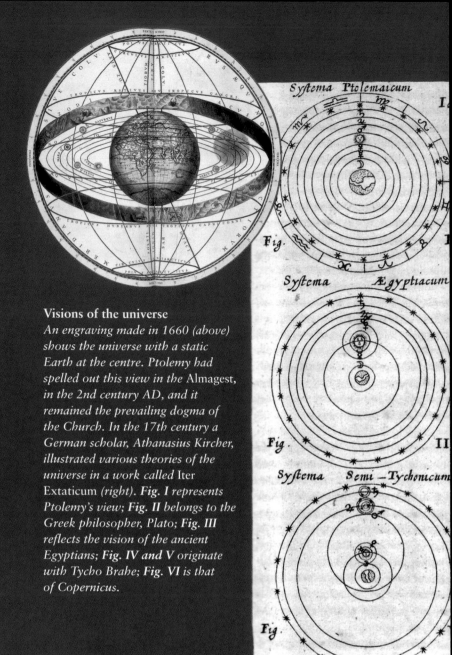

Visions of the universe
An engraving made in 1660 (above) shows the universe with a static Earth at the centre. Ptolemy had spelled out this view in the Almagest, *in the 2nd century AD, and it remained the prevailing dogma of the Church. In the 17th century a German scholar, Athanasius Kircher, illustrated various theories of the universe in a work called* Iter Extaticum *(right).* **Fig. I** *represents Ptolemy's view;* **Fig. II** *belongs to the Greek philosopher, Plato;* **Fig. III** *reflects the vision of the ancient Egyptians;* **Fig. IV and V** *originate with Tycho Brahe;* **Fig. VI** *is that of Copernicus.*

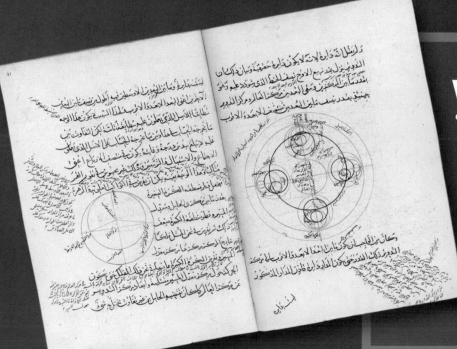

THE ARAB CONTRIBUTION

Was Copernicus directly influenced by Arab astronomers? Scholars have looked particularly closely at a possible link with Nasir ad-Din al-Tusi (1201–74). While not challenging the Ptolemaic view that the Earth lay at the centre of the universe, the Persian savant came up with a concept known as the 'Tusi couple' to explain the apparent movement of the planets on the celestial sphere. Copernicus apparently adopted this idea, but no evidence has yet come to light to show a direct chain of transmission from al-Tusi to Copernicus, so the question remains open.

Geometrical solutions *In his* Al-Tadhkira, *al-Tusi addressed problems of geometry, while also adding commentaries on Ptolemy's* Almagest.

supralunary rules applied to the heavens. But in other respects Copernicus went along with the established principles of Graeco-Arabic cosmology, accepting the idea of a world confined within a heavenly sphere in which the stars moved in regular, circular orbits.

A revolution of the talents

The Middle Ages in Europe have long been considered barbarous by post-Renaissance commentators, but in reality they were a time of great intellectual ferment. From the 12th century on, Latin translations of the works of Classical Greece and of works of Arabic learning spread across the Continent from Islamic Spain, then the hub of scholarly activity. From this melting pot of Classical, Arabic, Christian and Jewish cultures, the seeds of modern thought emerged. Among the pioneers who played a part in preparing the groundwork for the scientific revolution were

Copernican model
The Polish astronomer correctly deduced that the Earth turns on its own axis daily and revolves around the Sun over the course of a year. This illustration of his universe shows the Earth and the other planets revolving around the Sun. Less scientifically, it shows the signs of the zodiac around its rim.

AN UNWILLING REVOLUTIONARY

In his lifetime, Copernicus (1473-1543) was a cleric, physician, jurist, economist, diplomat and politician, who seemed little inclined to set off a revolution. He made time for his astronomical researches in between practising medicine, undertaking diplomatic missions, writing treatises on monetary reform and helping to defend Polish cities against assaults by the Teutonic Knights. Most of his productive work on astronomy was done in three periods. From 1491 to 1497 he studied at the Krakow Academy, where he made his first observations. He worked out the basis of his cosmic system between 1504 and 1514 and circulated his findings in a six-page manuscript called the *Commentariolus*. From 1530 on he prepared his masterwork, *De revolutionibus orbium coelestium* (right). In 1539 a German mathematician known as Rheticus travelled to Frombork, where Copernicus had lived since 1510, to persuade him to publish his findings. Eventually Copernicus agreed. Legend has it that as he drew near to death, in May 1543, he emerged from a coma just long enough to cast his eyes on the first printed copy of his work, brought by Rheticus from Germany.

An educational toy
This astronomical globe (below) was made in 1786 for the Comte d'Artois. A younger brother of France's King Louis XVI, the Comte would later inherit the throne as Charles X.

the Persian scholars al-Tusi and al-Farghani, the German astronomer Johannes Müller (also known as Regiomontanus), the French scholars Jean Buridan and Nicole Oresme, and the early English scientist and philosopher Roger Bacon, among many others. Even Christopher Columbus played a part by re-emphasising the idea – already familiar to the ancient Greeks and Arabs, but usually forgotten since – that the Earth was round. A century before Copernicus a German cleric, Nicholas of Cusa, had even dared to question the Ptolemaic hierachy, although in his case on theological grounds; he argued that God, and not the Earth, was the true centre of the universe.

Much still remained to be done when Copernicus passed away at Frombork in the north of Poland in 1543. By undermining the foundations of Aristotelian physics, he had left his fellow-scholars with unfinished business on their hands. There was no question of going back, for the great leap forward had already been taken. The only option was to press on towards a whole new image of the universe. The task would take a century and a half to accomplish, culminating in 1687 when Sir Isaac Newton published his theory of universal gravitation. This finally provided a definitive solution to problems that Copernicus had raised. At the risk of oversimplification, once Copernicus had revealed the Earth to be a planet like any other, terrestrial phenomena could no longer be explained in the light of the Earth's privileged position at the centre of the cosmos, as Aristotle had affirmed. In Newton's breakthrough, the force of gravity that ruled the movement of the Earth also acted in similar fashion on other heavenly bodies.

A new view of the universe

After Copernicus had quietly created the need for a new physics, a host of scholars dedicated their lives to illuminating the mysteries that he left behind as his legacy. One notable lesser-known figure was an Englishman named Sir Thomas Digges. An early proponent of Copernicus's theory, Digges went beyond the master's own heliocentric vision by proposing, in 1576, that the stars were not fixed on a celestial sphere but were instead scattered at differing distances across an infinite universe.

Giordano Bruno had also pushed ahead down the relativistic path indicated by Copernicus, treating the Sun as a star like any other. Galileo established mathematical laws to explain the motion of objects on Earth. He discovered the principle of inertia, by which bodies moving on a level surface continue in the same direction at a constant speed unless disturbed, and he also invented the first effective astronomical telescopes, making further discoveries possible. Johannes Kepler, a contemporary of Galileo, worked out the

Heavenly spheres
Dating back to Classical times, the first armillary spheres were built to show the motions of heavenly bodies around the Earth. This post-Copernican model has the Sun at its centre.

laws that govern the movements of the planets, demonstrating that their motion is not circular and uniform but elliptical and variable. He expressed his findings in mathematical formulae, providing a vital stepping stone for Newton.

It was left for Newton to finish off the Copernican revolution with his theory of universal gravitation. With his contribution the whole glorious celestial mechanism was finally laid bare. Newton acknowledged the role played by his predecessors in a celebrated remark, written in a letter to his fellow-scientist Robert Hooke: 'If I have seen further, it is by standing on the shoulders of giants.' Doubtless, he had Copernicus, Galileo and Kepler in mind as he wrote, but in fact a multitude of scholars had helped to prepare the groundwork for the discoveries by the great men of science. The Copernican revolution was in a very real sense a collective effort whose lasting legacy was nothing less than the rise of modern science.

Down into the ocean's depths

The hostile underwater environment long repelled any attempts to penetrate the secrets of the silent world beneath the sea. Even after early rudimentary submersibles showed it was possible to survive underwater, many technical problems had to be overcome before modern submarines enabled explorers to reach the ocean's depths. The motives driving innovation were mostly military.

Early endeavours
A submarine designed by Giovanni Borelli (right) was powered by oars with webbed tips; the Italian inventor was inspired by his observations of marine animals. The oars used in Cornelius Drebbel's earlier vessel (below), first tested successfully in 1620, were more conventionally shaped.

In 1620, King James I and thousands of Londoners observed a novel experiment taking place on – or rather, under – the Thames. The Dutch engineer Cornelius Drebbel had already spent some years in England when he began to experiment with an idea first outlined more than 40 years earlier by the mathematician William Bourne. A former sailor, Bourne had envisaged a covered and water-proofed craft capable of travelling underwater. Drebbel decided to put the idea to the test.

Essentially, the craft he came up with was a covered rowing boat made of wood and leather. The vessel was propelled by 12 oars and in order to submerge it employed a system of pigs' bladders beneath the rowers' seats that were connected by pipes to the water outside. By allowing the pigs's bladders to fill with water, the rowers reduced the buoyancy of the vessel until it sank; to resurface, they simply squeezed the water out again. The submarine must have had a rudimentary air-supply, but how it worked is a mystery today. Drebbel continued to experiment over the next few years, eventually building a craft that could stay submerged for 3 hours at a depth of 4m, managing a 4-mile underwater journey from Westminster to Greenwich.

Some years later the Italian scientist Giovanni Borelli invented a device that employed levers to fill and empty goatskin sacks that served as ballast. The French friar and polymath Marin Mersenne, best known now for his work on prime numbers, came up with an idea for an underwater vessel driven from the inside by a central paddle-wheel. An engineer named Jean Bérié tested out such a contraption at St Malo in 1647. For the rest of the century men continued to wrestle with the problems of propulsion, ventilation and water pressure that submarine exploration entailed.

From the *Turtle* to *Nautilus*

The next milestone was the *Turtle*, an egg-shaped wooden shell reinforced with steel bands, which was the brainchild of a Yale University student named David Bushnell. It was topped by a small, cylindrical turret equipped with thick glass portholes. To submerge the vessel, sea-water was allowed into a sealed chamber by operating a valve under the pilot's feet; to surface, the water was forced out with a hand-pump. To move forwards the operator turned a hand-cranked screw propeller.

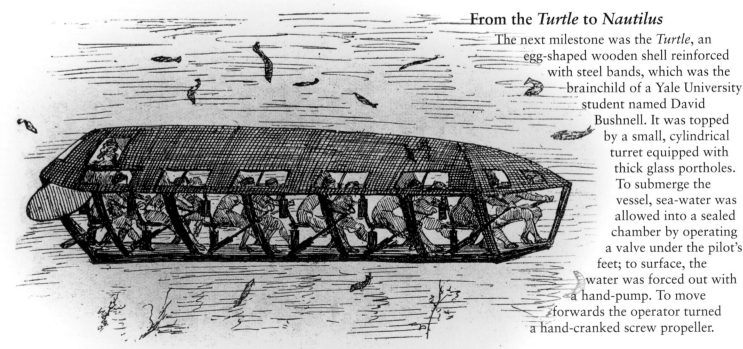

THE DIVING BELL – ECHOING AN ANCIENT DREAM

The earliest known device conceived for exploring underwater was the colimpha, a primitive diving bell supposedly invented for Alexander the Great in the 4th century BC. References to the colimpha in primary sources are unclear, but it probably consisted of a watertight barrel equipped with a transparent top that was dropped vertically into the water to prevent the air inside from escaping. Such a contraption could not have descended more than 3m below the surface before succumbing to water pressure. In Renaissance times Leonardo da Vinci produced a sketch of a diver being supplied with air through a tube. In 1553 an Italian military engineer named Francesco de Marchi explored two wrecks on the bottom of Lake Nemi near Rome using an iron-ringed wooden bell provided with holes for the diver's arms and legs.

The first truly operational diving bell was devised in 1690 by the astronomer Edmund Halley. It was a truncated wooden cone, covered in a casing of lead which stabilised it, and could carry two people down to a depth of 18m. A porthole provided light, while weighted casks of air dropped from the surface supplied ventilation. One man could venture out onto the seabed – to mend seawalls, perhaps, or to fix bridge pylons – wearing a diving helmet supplied with air through a tube from the bell.

At the end of the 18th century the English engineer John Smeaton went one step further, using cast iron as the principal material for his device. He also added a hand pump to draw air through a tube and installed an air reservoir and a check valve to prevent dirty air from backing up when the pump was not working. This was, in effect, the first modern diving bell.

In the years that followed, the introduction of diving suits revolutionised underwater exploration, but to this day diving bells have a use as decompression chambers. Divers can now work as deep as 180m below the waves carrying out essential seabed tasks or maintaining offshore rigs, then decompress in safety as they return to the surface in specially constructed vessels.

Alexander's colimpha
An engraving of 1850 (above), based on a 13th-century miniature, imagines the underwater 'barrel' supposedly invented for Alexander the Great.

Practical application
Halley's diving bell, designed in 1690, is shown here in a cutaway illustration from a 19th-century scientific encyclopaedia.

Underwater urn
A scale model of the Turtle, *the pioneering submersible that American forces tried to use against the British fleet in the War of Independence in 1776.*

The *Turtle* was a weapon of war, designed to place explosive charges on the hulls of British ships in the American War of Independence. Piloted by an army sergeant named Ezra Lee, it went into action against the British flagship HMS *Eagle* in New York harbour in September 1776, but failed in its mission.

Twenty-one years later another American, Robert Fulton, produced the *Nautilus,* an improved version of the *Turtle.* Fulton crossed the Atlantic to offer his services to Napoleon Bonaparte, who was then making plans to invade England. Bonaparte was sufficiently intrigued to provide Fulton with the

The *Turtle* in action
An artist's impression of the Turtle *going into action against HMS* Eagle *(below). Naval historians now believe that the failure to sink the ship stemmed from the pilot's inability to hold the submersible steady while he drilled a hole for explosives in the British flagship's hull.*

means to try out his project. Made entirely of metal, the machine housed three men to crank its propeller. It also sported a 4m collapsible mast with a small sail attached for travelling on the surface, making it the world's only wind-powered submarine. The *Nautilus* submerged for four hours off Brest in 1801. Its mines were also tested successfully, sinking a 12m sloop provided for the purpose by the French navy. But Napoleon remained unimpressed and cut off Fulton's funding. Some 70 years later *Nautilus* was remembered by Jules Verne, who gave the name to the submarine in his novel *Twenty Thousand Leagues under the Sea.*

Problems of propulsion

The first submarine to use mechanical power was the *Plongeur,* built to plans drawn up by the French marine engineer Charles Brun and first trialled in 1863. Powered by compressed air, the vessel carried a crew of seven and had a maximum dive time of two hours, but was taken out of service after nine years due to stability problems. By then a Spanish engineer, Narcisse Monturiol, had successfully tested the *Ictineo II*. This used an anaerobic propulsion system fired by the chemical reaction of zinc, potassium chlorate and manganese dioxide. Meanwhile, in Britain, Reverend George

A PRODUCTIVE CAREER

Robert Fulton, designer of the *Nautilus* (below), was born in Pennsylvania in 1765, the son of Irish emigrants. His father died when he was three years old, and the young Robert took a job with a Philadelphia jeweller before embarking, at the age of 17, on a career as a painter of miniatures. Moving to London to practice his trade, he soon abandoned his brushes to take up engineering.

Besides the *Nautilus*, Fulton's inventions included a machine for digging trenches as well as devices for cutting and polishing marble, for spinning flax and for twisting hemp. His greatest achievement was the development of steamboats. His paddlewheel-powered *North River Steamboat* entered commercial service on the Hudson River in 1807. He also designed a steam-powered warship for the US Navy that was launched shortly after his death in 1815.

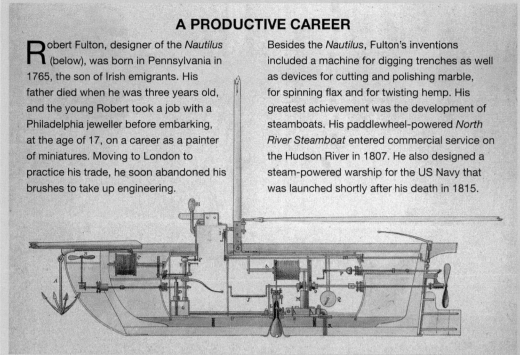

Coming up for air
The O'Higgins (below) one of two Scorpène class submarines delivered to the Chilean navy in 2005. The vessels, which were developed in a joint collaboration by French and Spanish companies, employ an air-independent propulsion system that allows them to remain submerged for three weeks or more at a time.

Garrett was working on a steam-powered submarine. A prototype, the *Resurgam I*, emerged in 1878; a much-improved version, *Resurgam II*, carried a crew of three and was successfully tested underwater in 1879.

The first practical submarines powered by an internal combustion engine were the brainchild of the Americans John Holland and Simon Lake. The *Holland VI* combined electric motors for submerged travel with gasoline engines for use on the surface; in 1900 the US Navy commissioned 12 of them. Lake's pioneering design, the *Argonaut I*, contained both a 'locked' chamber to hold the diver and a rudimentary 'viewing tube'.

By the 20th century most submarines were powered by diesel engines. The vessels acquired huge strategic importance in World War I, when German submarines armed with torpedoes sought to cut Britain's supply lines from overseas. From that time on undersea warfare became a crucial part of military strategy; in World War II, U-boats were central to German tactics in the Battle of the Atlantic. By the 1950s nuclear propulsion was enabling vessels to stay submerged for months, lurking 300m or more beneath the surface, ready to unleash an arsenal of ballistic missiles.

Auguste Piccard's bathyschaphes, which were operational between 1948 and 1982, were designed for exploring the ocean depths down as far as 11,000m. In the 1980s that role was taken over by deep-diving submersibles, which were used to investigate hydrothermal vents and the mysterious life-forms in the most profound and sunless depths.

The herald of a new world

Galileo brought together mathematics and physics, the Earth and the heavens. His novel approach to studying objects in motion, based on careful observation and measurement rather than abstract philosophy and religion, laid the foundations of modern mechanics and marked the end of Aristotelian physics. He paid for the audacity of his ideas with his liberty.

Portrait of a scientist
The Paduan artist Ottavio Leoni drew this likeness of Galileo around the year 1600.

Following his father's wishes, Galileo Galilei initially considered a career in medicine. But in 1583 he left Florence, where he had spent his adolescence, to return to his native town of Pisa and there he discovered mathematics. Falling in love with the beauty of the subject, Galileo neglected his medical studies in favour of Euclidian geometry. In that same year he became fascinated by the pendulum-like motion of a chandelier in Pisa's cathedral, noting how the radius of the arc gradually diminished with each swing, seemingly slowing down until the swaying finally stopped. Timing the oscillations against his own pulse, he realised that long and short swings both took the same amount of time. This observation led him to propose the theory (only approximately true) that pendulums are isochronous – that is, their back and forth motions occupy the same interval of time regardless of the distance covered. He then devised a mathematical formula to calculate the length of the swing. Galileo had taken the first step towards the creation of a new science: mechanics.

In 1585 he broke off his medical studies to return to Florence and teach mathematics. There he devoured the works of Archimedes, Plato, Pythagoras and Aristotle, regarding the latter in particular with a critical eye. He published a treatise called the *Bilancetta* ('Little Balance') on tipping points, then in 1589 he was offered the chair of mathematics at the University of Pisa.

A new way of thinking

In his new position Galileo continued to combine direct observation with deductive reasoning. He carried out experiments on falling bodies, including the famous one supposedly conducted from the Leaning Tower of Pisa (see panel below). In 1590 he published his first work on mechanics, entitled *De Motu* ('On Motion'), in which he claimed that 'anyone who does not understand motion does not understand nature'. While refraining from openly criticising Aristotelian physics, he

THE PISAN EXPERIMENT – FACT OR FICTION?

A biographer described how Galileo disproved Aristotle's notion that a heavy object falls faster than a lighter one by dropping pairs of balls of the same material but different mass from the Leaning Tower of Pisa. Two by two, each mismatched pair supposedly touched the ground at the same time. More recent historians prefer to see the tale as apocryphal. Rather than the experiment actually taking place, it seems more likely that Galileo merely suggested it as a possibility.

nonetheless managed to irk some of his university colleagues with his unconventional views. Eventually he lost his post – and for a time the favour of his Medici backers – by publicly doubting the efficacy of a dredging machine that had been invented by the son of the grand duke of Tuscany.

In 1592 Galileo once more left Pisa, this time to take up a position at the University of Padua, then in the territory of the Republic of Venice. He taught maths, astronomy, architecture and applied mechanics, enjoying an improved salary and greater freedom of expression. A radical in academic terms as well as in his scientific methods, he preferred to improvise his lectures, devoting some of the available time to practical experiments. His methods evidently worked, attracting thousands of students from all over Europe.

The 18 years that Galileo spent at Padua were the most productive of his life. He came up with a stream of observations, publications and inventions, among them a design for a horse-drawn water pump that he had patented and a 'geometric and military compass'. To test Aristotle's claim that objects naturally 'desired' to return to their place of origin, he experimentally rolled balls down slopes, and from his observations he drew the principle of inertia: that a body moving on a level surface will continue in the same direction at a constant speed unless disturbed. He also proposed that, in the absence of any significant resistance, falling bodies would fall with a uniform acceleration. His brilliant mind won him as many friends in high society as he made enemies at the university. Above all, he had at that time the support of the Jesuits and of the Venetian civil authorities.

Bringing the heavens closer

When his contemporary Giordano Bruno was burned at the stake for heresy in Rome in 1600, Galileo – probably already a convinced Copernican – remained silent. Four years later a supernova appeared in the sky and he tried, unsuccessfully, to measure its parallax. From this failure he knew that he must be observing a distant phenomenon and not – as Aristotle's worldview demanded – something occurring in the sublunary sphere. He now proceeded to criticise the accepted view of the universe in a series of lectures,

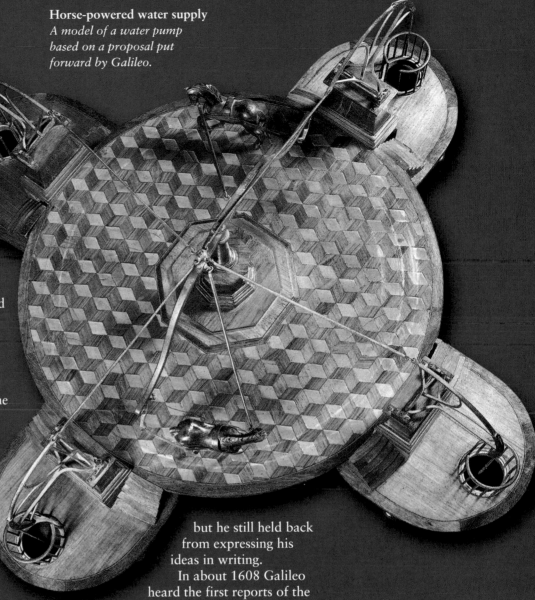

Horse-powered water supply
A model of a water pump based on a proposal put forward by Galileo.

but he still held back from expressing his ideas in writing.

In about 1608 Galileo heard the first reports of the telescope – a recent Dutch invention that could magnify objects three times. He quickly built models of his own, achieving first ninefold and then thirtyfold magnification. Using the new device he was able to observe sunspots, mountains and craters on the Moon, and the phases of Venus. He saw orbiting bodies around Saturn that would later be identified as the planet's rings. Above all, he became the first man to observe the four largest moons of Jupiter. None of these phenomena could possibly accord with Aristotle's concept of a flawless cosmos centred on the Earth.

Galileo outlined his findings in *Siderius Nuncius* ('The Starry Messenger'), published in 1610, and organised public demonstrations of his telescope. His academic rivals hastened to dismiss the telescope, claiming that it distorted the vision, but Galileo's reputation only continued to grow. The doge of Venice sought to reward him with a professorship and salary

THE BODY'S CLOCK

Having recognised and spelled out the isochronous nature of pendulum swings, Galileo went on to invent the first effective device for measuring the human pulse.

Galileo's instrument
Galileo used telescopes like this one (above) – shown in a reconstruction with the Duomo of Florence behind – to observe mountains and craters on the face of the Moon. He published his findings, including these two moon sketches (above right), in 1610 in The Starry Messenger.

for life, but he preferred to move to Florence, taking up an offer to be court philosopher and mathematician to the grand duke of Tuscany.

Defending Copernicus

Considering himself by now to be safe from his enemies, Galileo began to defend Copernicus. He thus laid himself open to accusations of heresy, but he was still protected by the Jesuits of the College of Rome, who accepted his astronomical observations, if not the conclusions he drew from them. In 1611 he was received into the prestigious Lincean Academy, where he established friendly relations with Cardinal Barberini, the future Pope Urban VIII. Then, in 1613, he published his *Letter on Sunspots,* and in so doing drew on himself the attention of the Inquisition.

Rather than challenging his science, his opponents preferred to base the case against him on theological grounds. Stubbornly, Galileo insisted that a literal interpretation

of the Bible with regard to the fixity of the Earth could only be mistaken; reason showed that the subject was a matter for science, not religion. Battle was joined.

In 1616 matters came to a head when the Church formally condemned Copernicus's heliocentric view of the universe, placing his work on the Vatican's Index of forbidden books. Galileo was summoned to Rome and his *Letter on Sunspots* was condemned. He still had influential friends willing to argue for him, but he now lost the support of the Jesuits.

Gagging free thought

Galileo response was to publish his *Dialogue concerning the Two Chief World Systems,* which took the form of a conversation between three men: on the one hand Simplicio, who defends Aristotle and Ptolemy, and on the other Salviati and Sagredo, both supporters of Copernicus's views. The book, which came out in Florence in 1632, was written in Italian rather than Latin to ensure that it reached the widest possible audience. In this work Galileo directly defied Church dogma. Not only had the publication not been authorised by Rome, but by coming down against Simplicio, it specifically contravened the 1616 injunction against promoting the Copernican system.

Even his old friend Barberini, now pope, could no longer save him. Galileo was convicted of heresy. In 1633 he was forced to publicly recant his ideas and was condemned to life imprisonment, subsequently commuted to perpetual house arrest.

FROM GALILEO TO EINSTEIN

In 1638, in his *Discourses and Mathematical Demonstrations relating to Two New Sciences,* Galileo suggested that a passenger in an enclosed cabin on board a boat at sea would have no way of knowing whether the vessel was static or moving over calm waters. The principle came to be known as Galilean relativity, and it laid the way for Einstein's future work. Einstein himself would one day write: 'The discovery and use of scientific reasoning by Galileo … marks the real beginning of physics.'

In his home near Florence Galileo devoted himself to preparing the *Discourses and Mathematical Demonstrations relating to Two New Sciences*, covering respectively kinematics (the study of motion) and the strength of materials. The book was eventually published in the Netherlands in 1638. In it he spelled out his law of falling bodies, preparing the way for Newton's later work on gravity.

Galileo lost his sight when he was 72 and died in 1642 at the age of 78. The Church's hostility pursued him long after his death. Although Copernicus's view that the Earth revolves around the Sun was accepted in 1822, it was 1992 before Pope John Paul II made a public apology for the way that Galileo had been treated by the Roman Catholic Church.

RAISING THE BAN

Most of Galileo's works were removed from the Vatican's Index of forbidden books in 1718, but the *Dialogue concerning the Two Chief World Systems* remained proscribed. Its uncensored publication was finally permitted only in 1835. Perhaps to make amends for such injustices, the Roman Catholic authorities proposed in 2008 to erect a statue of Galileo within the Vatican's walls.

Clock mechanism
A 19th-century model produced to Galileo's original pendulum design.

Professor at Padua
Painted by a Mexican artist, Felix Parra, in 1873, this portrait shows Galileo explaining his theories of the universe to a young monk attending his lectures at the University of Padua.

PADUA, CAPITAL OF ANATOMY

Harvey studied anatomy at Padua, a renowned centre of medicine where such luminaries as Vesalius, Renaldus Colombus and Fallopius had all taught. Harvey's master was Fabricius, the discoverer of venous valves and the man responsible for constructing the world's first anatomical amphitheatre.

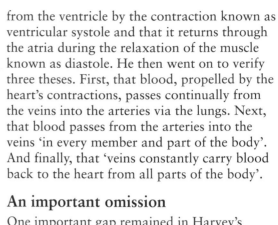

Veins and arteries
An illustration of the circulation of the blood, from the French Encyclopédie of 1762.

from the ventricle by the contraction known as ventricular systole and that it returns through the atria during the relaxation of the muscle known as diastole. He then went on to verify three theses. First, that blood, propelled by the heart's contractions, passes continually from the veins into the arteries via the lungs. Next, that blood passes from the arteries into the veins 'in every member and part of the body'. And finally, that 'veins constantly carry blood back to the heart from all parts of the body'.

An important omission

One important gap remained in Harvey's system: the link between the arteries and veins. In his great work, he regretted the fact that none of his experiments had revealed the connections, adding that he had spent entire nights and much effort looking as hard as he could without ever having managed to identify a single one. His intuition that such a link must exist was correct, but without access to a microscope it was beyond his grasp. The discovery of capillary vessels joining the two had to wait until 1661, when Marcello Malpighi made the breakthrough.

The system Harvey described was not perfect. There was still work to be done on the influence of the central nervous system on breathing, on the mechanism of the heart's contraction and on the oxygenation of the blood by gaseous exchange in the lungs. But the progress he had made propelled physiology to the rank of a science. It also demolished so much dogma it caused huge controversy: learned opinion in Europe divided henceforth into circulators and anti-circulators.

THE FIRST BLOOD TRANSFUSIONS

William Harvey's discoveries directly inspired the first experiments with blood transfusion. In 1666 a Cornish doctor named Richard Lower bled a medium-sized dog almost to death, then refilled its veins with fresh blood taken from a mastiff. The first dog recovered, but the mastiff died. The following year one of Louis XIV's court physicians, Jean-Baptiste Denys, injected calf's blood into an insane patient in the hope of calming him. After a third transfusion the patient died; Denys was charged with murder but acquitted. The treatment was subsequently banned, and 150 years would pass before the practice of transfusion was revived.

Life's blood *An illustration from the* Armamentarium Chirurgicum, *a textbook on surgery by Johannes Scultetus (1595-1645), shows blood taken from a dog being transfused into a patient suffering a haemorrhage.*

Harvey and the King
A painting of 1848 by Robert Hannah shows William Harvey demonstrating his theory of the circulation of the blood to Charles I.

Conflicting opinions

Rivals rushed to mock Harvey, making much use of the word *circulator*, which in Latin meant 'pedlar' or 'charlatan'. The French anatomist Jean Riolan wrote to him to insist that he had made many foolish mistakes and errors. Most controversial of all was the idea that the blood flowed in a circular motion. As late as 1670, the dean of the Paris Faculty of Medicine insisted that the notion was 'the childish fancy of a lazy mind, a mirage embraced by Ixions to bring forth centaurs and monsters'. Even René Descartes resisted the notion of autonomous cardiac contractions, preferring to believe that the expansion of the ventricles was caused when blood was abruptly vaporised by the heart's heat.

Yet in the end Harvey was proved right, when Malpighi's work with microscopes confirmed the accuracy of his observations. Although his discovery did not deal an instant death-blow to the Galenic system – indeed, Galen's ideas survived for many years, most notably in the practice of blood-letting – he had delivered a fatal wound, opening the door to a new physiology based on direct observation rather than received opinion.

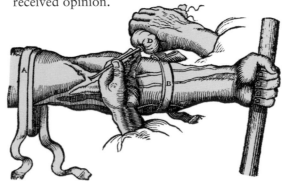

Observing veins and venous valves
This illustration from Harvey's work shows how he used ligatures to control the flow of blood to make veins stand out for observation in the lower arm.

43

From matchlock to flintlock

Following an ordinance by France's King Henri IV, an inventor from Normandy drew on the experience of 150 years of firearms use to transform the musket into an effective all-weather weapon. He came up with the flintlock musket – simple, cheap and easy to handle, it was eventually adopted by armies across the world.

Firing early muskets
The first muskets were heavy, so soldiers usually employed a fork-rest to fire them. Later models had longer stocks and curved butts, enabling them to be fired from the shoulder.

match

flash pan

Firing a matchlock
To fire a matchlock musket, the shooter put priming powder in the flash pan and then ignited a slow-burning match attached to a curved lever. When the trigger was pulled, the match was lowered into the pan, setting off sparks that fired the main charge in the gun's barrel.

Guns in transition
The illustrations here show three early firing mechanisms: the earliest is the matchlock mechanism (far left), followed by the wheel-lock musket of 1610 (above) and finally that of the flintlock musket, made some years later.

Marin Le Bourgeoys lived at Lisieux in Normandy, where his father was a locksmith and clockmaker and his brother a gunmaker who specialised in producing arquebuses. The young Marin grew up with a passion both for precision engineering and for arms, but at the time there were strict rules in France as to who could work on the various parts of weapons. For instance, arquebusiers were not allowed to make the barrels of their own weapons – that was the exclusive monopoly of blacksmiths. Only clockmakers were licensed to provide triggers. Another specialist group of craftsmen produced nothing but gun-butts.

This situation changed in 1608, when Henri IV passed an ordinance exempting court artisans from the restrictive guild regulations and creating the new profession of armourer. Marin Le Bourgeoys had already passed his apprenticeship making triggers in his father's workshop. Suddenly, he could widen his horizons and he took advantage of the opportunity to develop a weapon that would prefigure the modern rifle.

The first portable firearms

The first firearms had reached Europe 150 years earlier, perhaps imported from Arab lands. The arquebus was effective, but it had one major drawback: it weighed 15–20kg and took two men to fire it, perched on a metal fork-rest rammed into the earth. Subsequently the design was improved by the introduction of matchlocks, which lowered a burning slow-match into the flash pan where the priming powder was set, and the guns got a new name: muskets. But they, too, had a serious weakness, as in damp weather conditions the powder would often fail to light.

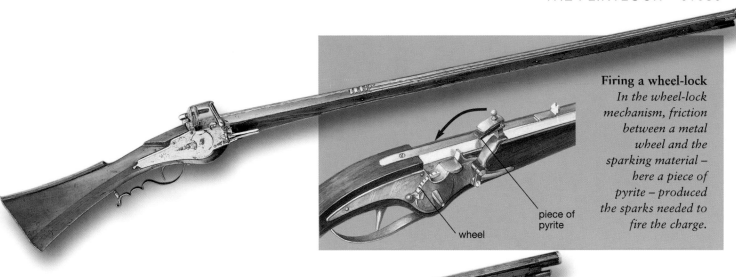

Firing a wheel-lock
In the wheel-lock mechanism, friction between a metal wheel and the sparking material – here a piece of pyrite – produced the sparks needed to fire the charge.

piece of pyrite

wheel

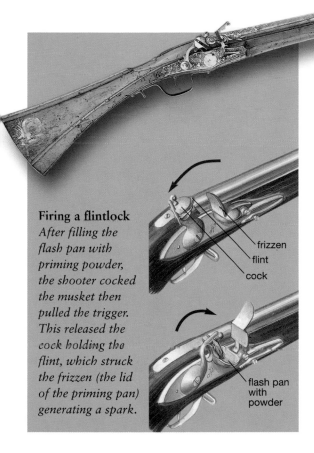

Firing a flintlock
After filling the flash pan with priming powder, the shooter cocked the musket then pulled the trigger. This released the cock holding the flint, which struck the frizzen (the lid of the priming pan) generating a spark.

frizzen
flint
cock

flash pan with powder

MORE THAN A FLASH IN THE PAN

In service for almost two centuries, flintlock muskets made a lasting contribution to the English language. The phrase 'going off at half-cock' recalls what happened when a gun was not properly cocked for firing: inevitable failure. 'Lock, stock and barrel' described a gun in its entirety, from butt to muzzle, while a 'flash in the pan' was what happened when the priming powder fired without igniting the main charge – the resulting flare-up soon fizzled out.

The next step forward was the invention of the wheel-lock, which employed a rotating metal wheel to produce a spark as it rubbed against a small piece of iron pyrite. Guns could henceforth be fired in any weather and the wheel-lock rifle was born. But pyrite was hard to come by, so at some point early in the 17th century, Marin Le Bourgeoys developed an alternative mechanism employing flint, which was common. The flint was attached to a pivoting hammer, which had to be set in the full-cock position for the gun to be fired; set at half-cock, it was in safety mode and would not ignite the charge. When the trigger was released, the hammer flew forward and the flint struck the frizzen – a piece of steel on the lid of the flash pan – which opened to expose the priming powder. The accompanying shower of sparks duly ignited the powder, which in turn sent flames shooting through a tiny touch hole connecting the pan to the main charge in the barrel and so detonated the shot.

The new weapon was cheaper and easier to produce than the arquebus; it was also shorter (at about 1.3m) and easier to handle. Another advantage was that it took less time to load. Impressed by the efficiency of Le Bourgeoys' flintlocks, Henri IV decided to keep their production a state secret. Perhaps as a consolation, he gave Le Bougeoys himself the honour of becoming the first court-appointed master-armourer.

Slow progress

Partly as a result of Henri's secrecy, armies elsewhere continued to use matchlock firearms throughout the first half of the 17th century. But word gradually spread. Infantry forces at that time were made up of pikemen and musketeers, and it was the former, tiring of

THE COMING OF THE MUSKET

Unlike the earlier, inefficient arquebuses, muskets helped revolutionise warfare. Musketeers had a greater range than bowmen – and, in addition, they could be trained in weeks, while good archers had to put in a lifetime of practice.

Battle scene
By 1855, when this encounter in the Crimean War took place (right), rifles were already equipped with bayonets for close-range combat.

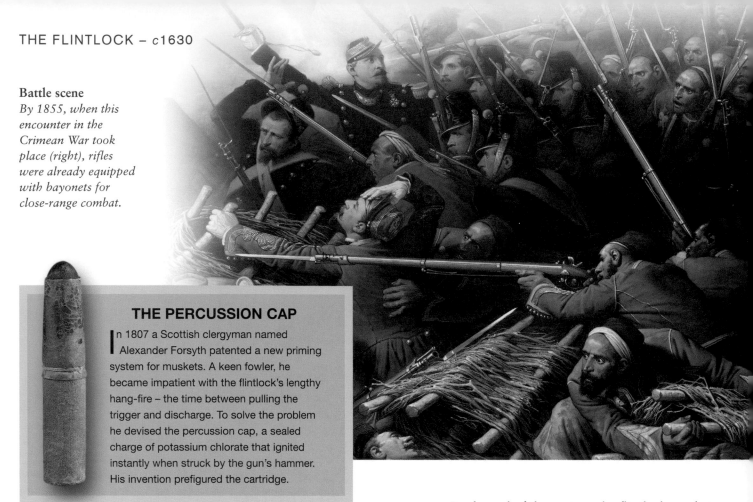

THE PERCUSSION CAP

In 1807 a Scottish clergyman named Alexander Forsyth patented a new priming system for muskets. A keen fowler, he became impatient with the flintlock's lengthy hang-fire – the time between pulling the trigger and discharge. To solve the problem he devised the percussion cap, a sealed charge of potassium chlorate that ignited instantly when struck by the gun's hammer. His invention prefigured the cartridge.

always playing second fiddle, who eventually began to demand the new flintlocks. Special corps of fusiliers – so called because flintlock muskets were known in France as *fusils* – began to be deployed. In Britain the guns were known in the Civil War years as firelocks.

The invention, late in the 17th century, of 'plug' bayonets that could be fitted into sockets on musket barrels sounded the deathknell for both pikemen and the matchlock musket. In future, musketeers had the wherewithal to fight both at a distance and at close quarters.

By the end of the century the flintlock musket reigned supreme. Yet as time passed, soldiers became aware that it, too, had its defects. Each shot was announced by a puff of smoke, revealing the shooter's position to the enemy. The guns were muzzle-loaded, and the barrels easily became dirty. Even the most efficient operator could only fire three or four shots a minute – more than Le Bourgeoys could ever have hoped for, but military ambitions had grown since his day. In the early 19th century, with the invention of breech-loading and the cartridge, the flintlock became outdated. The modern rifle was about to be invented.

Trench gun
During World War I, the French gun manufacturer Lebel adapted their 8mm calibre rifle to give some protection to soldiers firing from the trenches. The weapons went out of production in 1920.

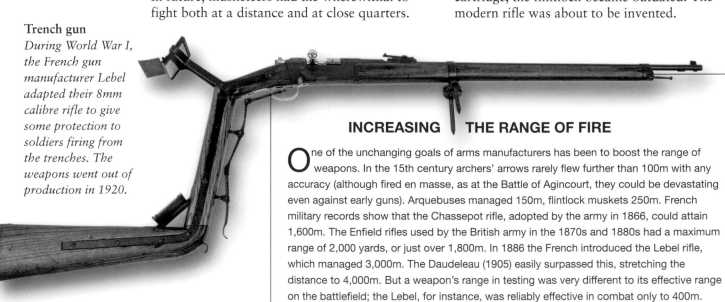

INCREASING ▼ THE RANGE OF FIRE

One of the unchanging goals of arms manufacturers has been to boost the range of weapons. In the 15th century archers' arrows rarely flew further than 100m with any accuracy (although fired en masse, as at the Battle of Agincourt, they could be devastating even against early guns). Arquebuses managed 150m, flintlock muskets 250m. French military records show that the Chassepot rifle, adopted by the army in 1866, could attain 1,600m. The Enfield rifles used by the British army in the 1870s and 1880s had a maximum range of 2,000 yards, or just over 1,800m. In 1886 the French introduced the Lebel rifle, which managed 3,000m. The Daudeleau (1905) easily surpassed this, stretching the distance to 4,000m. But a weapon's range in testing was very different to its effective range on the battlefield; the Lebel, for instance, was reliably effective in combat only to 400m.

The vernier calliper 1631

The vernier calliper is a precision tool that allows for highly accurate measurement. It was first developed in 1631 by the French mathematician Pierre Vernier. His starting-point was the classic sliding calliper, invented by the Chinese some 1,600 years earlier, which featured two jaws attached along a central ruler marked with graduated calibrations; one jaw was fixed to the end of the scale, the other moved to give the measurement. Vernier's innovation was to add a smaller scale that refined the result by a factor of 10. This new scale had nine separate equidistant markings designed to align with the fixed scale once in every 10 gradations. By reading off the point of alignment on the smaller scale, the user could obtain an exact measurement, not just in centimetres and millimetres but in tenths of a millimetre. More than two centuries later, the American engineer Joseph Brown produced a modern version of Vernier's calliper capable of measuring to a thousandth of an inch or a hundredth of a millimetre.

Precision reading
The scale on this calliper gives an exact measurement of the diameter of a ball bearing.

The umbrella 1637

Umbrellas reached Europe from the Orient in the 17th century. The Englishman Thomas Coryat described them in 'Crudities', an account of his travels in Italy published in 1611. Made of leather, their purpose was more to protect against sun than rain. French court records for 1637 list 'three umbrellas of oilcoth trimmed with gold and silver lace' among the personal effects of Louis XIII. It was 1680 before the word found its way into a dictionary, which noted that 'some women are beginning to use the term, although … laughingly'. In England, despite the climate, umbrellas were long considered effeminate, but successive improvements eventually made them both easier to use and socially acceptable. The first man in London to carry one regularly, from 1750, was Jonas Hanway. Henry Holland of Birmingham patented the first ribs made from steel tubes in 1840. Telescoping steel shafts date from 1934, invented by a German named Hans Haupt. New models henceforth had ribs designed to fold in the middle. Before long a push-button opening mechanism had been added and the modern, collapsible, convenient umbrella was born.

PRECURSORS

Devices for protecting the well-to-do from sun or rain date back a long way – to as far as 3,000 years ago in Mesopotamia (modern Iraq). Early models were made of palm leaves, stretched papyrus or peacock feathers, and were so heavy that porters were required to carry them. Chinese prints dating back to the 2nd century BC record waxed-paper parasols provided with pliable ribs.

Ineffective cover
A French cartoon from the early 19th century illustrates some of the problems of umbrella technology.

A DISTANT ANCESTOR?

In 1901 a sponge diver working off the Greek island of Antikythera discovered a strange device in an ancient wreck lying some 42m below the water's surface. Corroded and in pieces, it was obviously no longer in working order, but it was possible to make out pointers, dials, toothed wheels and illegible inscriptions. Scientists who studied it were sufficiently intrigued to try to make a replica, finally producing a model partly thanks to the use of X-ray techniques. In 2006 they revealed that the machine had a sophisticated gearing mechanism to indicate the position of the Sun and Moon in the sky and to predict eclipses. Two years later they found that it also served as a calendar, perhaps specifically for pinpointing the dates of upcoming Olympic Games. The device, which was dated to the second half of the 2nd century BC, is now thought to have originated not in Rhodes, like the ship it was found on, but in Syracuse in Sicily, the home town of Archimedes, raising the intriguing possibility that the great Greek scientist invented a calculator 18 centuries before Pascal.

The Antikythera mechanism
The device, shown here as it was originally found in 1901 (top left), was engraved with astronomical inscriptions. X-ray imaging (top right) revealed 30 bronze gears. This modern reconstruction (right) shows how it might have looked when first built.

without the drums and simply shown the results on the same axis as the dials, but that would have involved turning the box over to view the display windows, which would have been inconvenient. Given that his aim was to create a device that was easy to use, he was happy to introduce extra gearing in order to achieve that end. Ahead of his time as always, Pascal showed a firm grip of the discipline that would one day become known as ergonomics.

Having dialled in '19', the operator then had to add the 1, at which point the single-digit display reverted to zero. Pascal's innovation lay in devising a workable mechanism for carrying figures over from one column to the next, in this particular sum from the single digits to the tens. A system of gears linked the wheels. Each time a dial was turned a notch, it raised a tine that finally fell under its own weight once the full complement of 10 was reached, moving the neighbouring wheel on by one unit as it did so. In our sum, once the single-digit wheel returned to '0', the figure '1' in the display window of the tens dial automatically changed to '2'. Relying as it did on gravity, the calculator had to be placed flat to work properly. To prevent the machine from jamming, Pascal had the weight of connecting mechanisms diminish as the numbers got higher, with the heaviest tine – the one linking single digits to tens – driving all the rest.

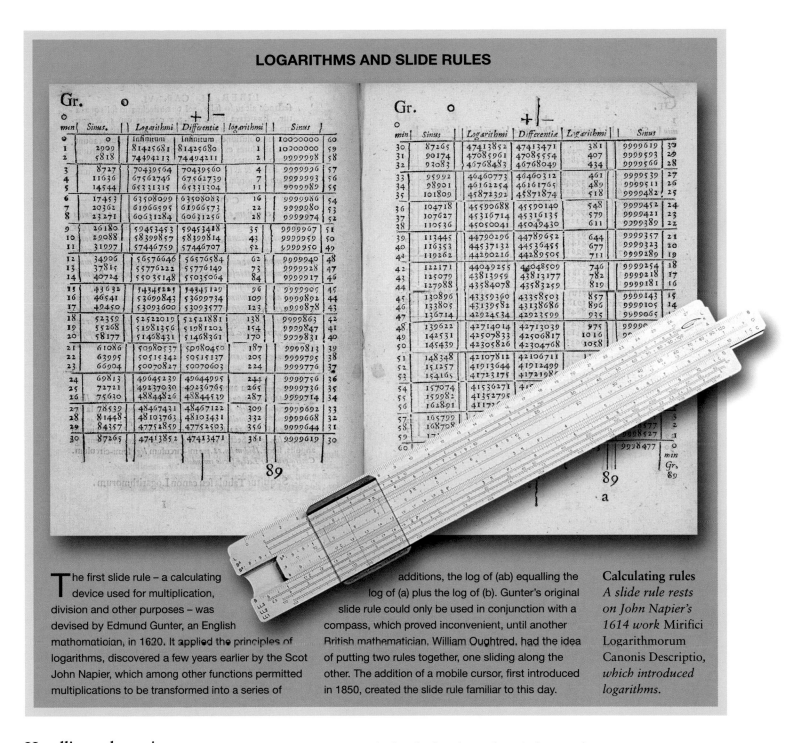

LOGARITHMS AND SLIDE RULES

The first slide rule – a calculating device used for multiplication, division and other purposes – was devised by Edmund Gunter, an English mathematician, in 1620. It applied the principles of logarithms, discovered a few years earlier by the Scot John Napier, which among other functions permitted multiplications to be transformed into a series of additions, the log of (ab) equalling the log of (a) plus the log of (b). Gunter's original slide rule could only be used in conjunction with a compass, which proved inconvenient, until another British mathematician, William Oughtred, had the idea of putting two rules together, one sliding along the other. The addition of a mobile cursor, first introduced in 1850, created the slide rule familiar to this day.

Calculating rules
A slide rule rests on John Napier's 1614 work Mirifici Logarithmorum Canonis Descriptio, *which introduced logarithms.*

Handling subtraction

Pascal's calculator could also be used for subtraction, inverting the process of addition. Take the sum 25–3. The first step was to manipulate a wooden rule that hid the base of the display windows to reveal the lower part of the revolving drums: the numbers here were in descending order rather than ascending. Using the stylus as before, the operator dialled up '25', then turned the single-digit dial three notches back: the display clicked back through '24' and '23' to '22' – the correct answer.

Having devised the prototype machine, Pascal now looked around for a craftsman capable of building multiple examples. He first approached a clockmaker, who tried to steal his invention, producing unauthorised copies. Pascal sought redress and in 1649 obtained a letter of patent signed by the youthful King Louis XIV, then just 11 years old, giving him the exclusive right to produce and sell the instrument, which henceforth became known as the 'pascaline'.

In its day the calculator was greeted with astonishment and admiration. For the first time, human ingenuity had come up with a device that could mechanically replicate the process of thinking. In her posthumous biography of her brother, Pascal's elder sister Gilberte would even claim that his invention

RENÉ DESCARTES 1596–1650
The idealistic materialist

The French philosopher René Descartes broke away from the certainties of medieval thinking. He put doubt at the very foundation of his belief system and created a 'universal science' based purely on reason, paving the way for the modern era. Yet Descartes personified the contradictions of 17th-century thought as well as its glories.

Face of a philosopher
A portrait painted by Frans Hals in about 1649, towards the end of Descartes's life.

Early in the 17th century, while a student at the Royal College in the small town of La Flèche, northwest of Tours in France, the teenage René Descartes was growing bored. His enquiring mind chafed at the school lessons he received there from the Jesuit priests. Philosophy, theology, physics – all were based on the ancient Aristotelian tradition and the subsequent commentaries of St Thomas Aquinas. Descartes would later compare their theories to 'proud and magnificent palaces that were built on sand and mud'. They could offer neither certainty nor a solid foundation for knowledge – in short, nothing that could be verified. Only mathematics, it seemed to him, met that criterion and rested on indisputable truths. His growing dissatisfaction would breed an ambition to find a key to the laws of nature, in the process setting off a seismic shift in the history of thought whose vibrations are still felt today.

Mystical illumination

René had been born in 1596 in the small town of La Haye in the Touraine region, which would later be renamed Descartes in his honour. Like his younger contemporary Blaise Pascal, he was a child of the educated provincial middle class and his mother died young – in Descartes's case when he was just one year old. His father was a counsellor of the provincial parliament in the adjoining region of Brittany. René was a delicate youth, but nonetheless decided to join up for military service at the age of 20. He developed an aptitude for fencing, fighting at least one known duel, and is thought to have written a treatise on the sport, now lost.

By his own account, three dreams came to him on the night of 10 November, 1619, that changed his life and with it the course of science. He was in winter quarters by the Danube in southern Germany, keeping warm in a room that he described as being like 'a hot stove'. First he had two nightmares involving swirling winds and thunder, but then, in the final dream, he found himself presented with two books – one an encyclopaedia, the other an anthology of poetry. Describing the experience soon afterwards, Descartes wrote that 'I glimpsed I know not what light, whose help can, I believe, dissipate the deepest gloom'. 'Filled with enthusiasm', he conceived the 'foundation of a wonderful science' that would bring together all human knowledge and wisdom. So in a curious irony, a moment of mystical illumination lies at the root of modern scientific rationalism.

The Dutch period

Descartes subsequently gave up his military career and for a time he travelled, before settling in Paris where he frequented salons and kept fashionable society. He also made notes for a book not published in his lifetime, the *Rules for the Direction of the Mind*.

By 1628 he was ready to devote himself full-time to his researches. He moved to the Netherlands, where he lived in seclusion for the next 20 years, following the motto *Bene vixit, qui bene latuit* – 'He lives well who lives well hidden'. He worked there on a *Treatise on the World*, which took a heliocentric view of the universe, but decided against publishing it when he heard of Galileo's condemnation by the Catholic Church in 1633.

In 1637 Descartes's most famous work, the *Discourse on Method* – to give it its full title, the *Discourse on the Method of Rightly Conducting the Reason and of Searching for Truth in the Sciences* – was published in Leiden. The book was written in French rather than Latin, the language of scholarship. The *Metaphysical Meditations* followed in 1641, then the *Principles of Philosophy* in 1644.

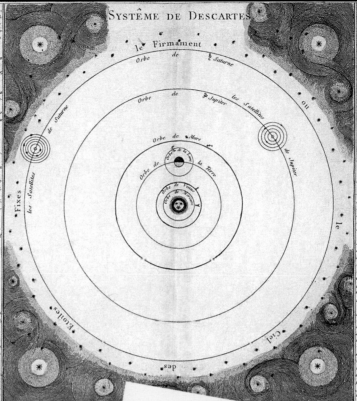

Scepticism and reason

What Descartes brought that was new to philosophy and science was a methodical scepticism and a determination to use reason alone to attain knowledge. To distinguish truth from falsehood and indisputable fact from unverifiable speculation, everything had to be sieved through a fine mesh of doubt. He made it his mantra 'never to accept anything as true that I did not clearly and distinctly know to be so'. Today such a maxim seems like simple good sense, but at the time its consequences were explosive, first in academic circles and later in the religious and political worlds. They have continued to resonate down the ages to the present day.

Descartes's uncompromising rationalism shook the principle of traditional authority to its foundations. For the philosopher and his followers, who became known as Cartesians, whatever was contrary to logic simply did not exist. Mathematics, with its infallible logic, became the pinnacle of natural truth. The ultimate task, then, was to unveil a *mathesis universalis* – a universal mathematics. For Cartesians this 'admirable science' would make humankind 'the master and proprietor of the natural world'.

Descartes believed that his way of viewing the world was applicable in every field of knowledge: 'All philosophy is a tree whose roots are metaphysics and whose trunk is

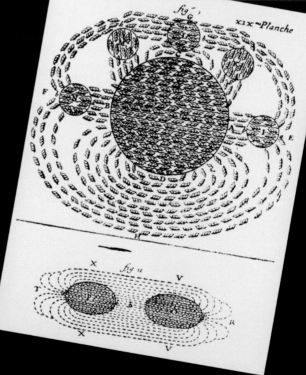

physics. The branches that spring from that trunk are all the other sciences, which can be reduced to three principal categories: medicine, mechanics and morals.' The method he spelled out opened the way to metaphysical reflection founded on doubt: if I question things, I am thinking, and if I am thinking, then I know that I exist. This was the basis of his famous formula: '*Cogito, ergo sum*' – 'I think, therefore I am'.

Descartes's cosmology
Descartes accepted the Copernican view that the Earth revolved around the Sun, but went further than Copernicus in viewing the Sun as only one star among countless others. He believed that planets spun in a vortex of light-bearing ether, a theory that was soon to be invalidated by Sir Isaac Newton.

As far as Descartes was concerned, his approach did not question the existence of God: in his view, if 2 plus 2 equalled 4, then some entity must have made it so. But it did affirm the existence of a thinking 'I', free to examine whatever a person wished.

Descartes at the Swedish court
A work by the artist Louis-Michel Dumesnil, painted after Descartes's death, shows the great philosopher (bottom right) demonstrating a problem of geometry to Queen Christina in the presence of the Prince du Condé, the philosopher Père Mersenne, and the Princess Palatine Elizabeth of Bohemia.

The Cartesian inheritance

In 1649 Descartes accepted an invitation from Queen Christina of Sweden to work at her court. He barely had time to publish his *Treatise on the Passions* before he died of pneumonia on 11 February, 1650. His remains were returned to France in 1667.

After his death Descartes attracted a school of followers – the Cartesians. They included sceptics like Malebranche, Leibniz, Spinoza and Hegel (who hailed Descartes as a 'hero of the realm of thought') and also men of faith, among them Bossuet, Fénelon and Pascal. Cartesianism never truly became a doctrine, and the Cartesians themselves had little in common except a respect for their master's essential principles: belief in the authority of reason in face of prejudice, and in rigorous logic and conceptual clarity as the foundation of all philosophic and scientific thought.

120 PRINCIPIORUM PHILOSOPHIÆ

THE RULE OF REASON

Descartes's rigorous use of reason acquired semi-mythic status in his native France, where Cartesian rationalism was raised almost to the level of a national characteristic. Yet however liberating his views may have been in their day, encouraging a sceptical attitude towards received wisdom, they eventually set into something resembling a dogma. In particular, the philosopher's dualistic vision of an immaterial mind trapped in a machine-like body ruled out holistic thinking about the mind-body connection. Similarly, his conviction that only humans possessed a mind led him to deny that other animals could experience either emotion or pain, thereby helping to legitimise the practice of vivisection and other human cruelties.

PARS TERTIA. 121

Principles of Philosophy
Published in 1644, Descartes' Principia Philosophiae *aimed to provide a firm foundation for philosophical reasoning.*

The body machine
For Descartes, reason was a function of an immaterial mind set in a corporeal brain. It was what distinguished humans from animals and automata, like this 16th-century archer.

Descartes the scientist
An illustration from the Treatise on Man *(left) demonstrates the relationship between sensory perception and muscular action.*

DESCARTES'S SCIENTIFIC LEGACY

Descartes left an important legacy in science, particularly in mathematics. In algebra he introduced the convention of using superscript to denote exponents, representing, for example, 2 squared as 2^2. More importantly, he is credited with laying the foundations of analytical geometry in an appendix to the *Discourse on Method*, where he outlined an approach for applying the principles of algebra to the study of geometry. His stated aim was to find a way of solving problems 'employing only circles and straight lines'. Legend has it that he was watching a fly on a window pane when he had the idea of locating the position of a point by the use of a grid and numbers, employing two separate figures – the ordinate and the abscissa – to pinpoint a location vertically and horizontally. These figures, as used for example in map references, are still known as Cartesian coordinates. In the field of optics Descartes formulated the law of refraction, known alternatively as Snell's or Descartes's Law – the two men discovered it

independently of one another. Other aspects of Descartes's work were quickly challenged. In astronomy, his view that planets were moved by disturbances in an invisible luminiferous ether was soon disproved by Isaac Newton. In medicine and physiology, his mechanistic concept of living organisms gave way in the 20th century to the psychosomatic view that the mind can influence the body. Trapped within the limitations of his own convictions, Descartes proved subject to error, not least in rejecting as contrary to logic Harvey's discovery of the contractions of the heart. His real contribution lay in giving priority to reason over established dogma and so opening up to question all accepted knowledge and conclusions – a step crucial enough to split the history of Western ideas into pre and post-Cartesian eras. It was onl in the 20th century, with the philosopher Ludwig Wittgenstein's assault on the limitations of formal logic, that Descartes's rigid rationalism came to be truly challenged.

Measuring air pressure

The invention of the barometer by Italian scientist Evangelista Torricelli marked a decisive breakthrough not just in the history of science but also in our conception of the physical world around us. He demonstrated convincingly the existence of air pressure and also contradicted the dogma that nature abhors a vacuum.

Torricelli's breakthrough
An engraving replicates the experiment that led Torricelli to invent the mercury barometer. The Italian scientist also formulated a law, still known by his name, that relates the speed at which a fluid flows out of an opening to the height of the fluid above the vent.

In Florence in 1635, the garden designers employed by Ferdinand II de Medici, grand duke of Tuscany, could not get the effect they wanted from the palace fountains. Despite the huge mass of water in the city's reservoirs, located below the level of the gardens, the height of the fountain jets never exceeded 10m, and they had been hoping for at least 12m or more. They turned for help to an elderly man who was living under house arrest nearby. That man was Galileo, imprisoned as punishment for his supposedly heretical writings. When he died in 1642, Galileo still had not fully resolved the problem, but he had sketched out a working hypothesis: that air has weight. The principle was subsequently teased out by his disciple Evangelista Torricelli, who also devised an instrument for measuring the pressure that air exerted – the barometer.

Getting water to rise

The subject was of far more than local interest, for water distribution in cities was a hot topic at the time. There was no known way of raising water uphill from a low-lying reservoir, other than using a mill or an Archimedes' screw. The reason behind this lack remained something of a mystery. If a siphon with a curved length of piping rising into the air was placed at the bottom of a reservoir containing thousands of tonnes of water, it seemed logical that the water should be driven up the tube with a force proportionate to the amount of water in the reservoir. Yet this did not happen:

in practice the water never rose higher than 10.3m, even if the air was removed from the tube. Galileo postulated that beyond 10.3m the force driving the water up was simply no longer sufficient to overcome the downward pressure of air on the column, bringing its

Iconismus X. pag. 28.

WEIGHT OR PRESSURE – IT'S ALL THE SAME

Measuring the weight of the air and air pressure are two sides of the same coin, rather like buying goods by the portion (say, 250g of fillet steak for £5) or by weight (at £20 per kilo). The price of steaks weighing 200g and 300g would obviously be different, but the cost per kilo would stay the same. Similarly with air pressure, the weight pressing on a given surface of, say, water or mercury depends on the area exposed; the bigger the area, the more air pressing down upon it. Yet whatever the surface area, the weight per square centimetre remains the same, and that is the air pressure.

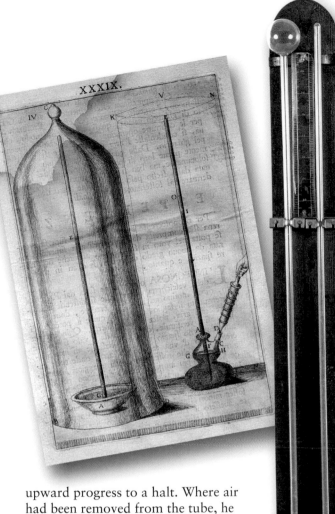

XXXIX.

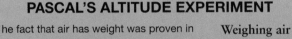

column of mercury should be 10.3m divided by 13.6 – just under 76cm. In his experiment, carried out in 1643, the level of the mercury stabilised at exactly the point he had predicted. The mercury that remained in the inverted tube created a vacuum above it, and thus Torricelli became the first scientist to create a sustained vacuum. This discovery was a milestone in physics, helping to pave the way for advances ranging from the development of steam engines to the creation of vacuums by particle accelerators.

In subsequent trials Torricelli noticed that the mercury level varied slightly from day to day, and that a drop in air pressure often presaged bad weather. No-one realised at the time that the air pressure was going up and down with the prevailing meteorological conditions, but Torricelli's observation at least opened the door to the development of later weather-forecasting methods.

Mercury in a tube
An early mercury barometer (left) and the principle behind how it works, as illustrated in an engraving dating from Torricelli's day (far left).

upward progress to a halt. Where air had been removed from the tube, he attributed this paradox to the limit of the water column's attraction to empty space, citing in his 1638 *Discourse concerning Two New Sciences* 'nature's famous abhorrence of a vacuum'.

Putting mercury to use

Addressing the problem in his turn, Torricelli took up Galileo's notion of the weight of air. He sensed that the solution must be related to the pressure that air exerted on the surface of the water in the reservoir. He viewed the system formed by the reservoir and the siphon piping as a balancing act between on the one hand the air pressing down on the reservoir's exposed surface and on the other the weight of the column of water in the pipe; equilibrium would be reached when the two balanced out.

To test out his theory, Torricelli filled a thin glass tube with mercury and upended it in a dish. He used mercury rather than water because of its greater density: 1 litre of mercury weighs the same as 13.6 litres of water. If his hypothesis was correct – that the pressure of air was the operative force – he would need 13.6 times less mercury than water in the tube to reach the point of balance, and therefore needed a much shorter tube. He calculated that the height reached by the

PASCAL'S ALTITUDE EXPERIMENT

The fact that air has weight was proven in 1648 by Blaise Pascal, who had a new-fangled barometer carried up the Puy de Dôme, a peak in the Auvergne region of France. On the summit, at 1,464m, the mercury levelled off at 63cm, compared to 76cm at the base of the mountain. This indicated that the weight of the column of air pressing down on the reservoir of mercury in the barometer was greater at lower altitude. The higher up the mountain the barometer went, the shorter the mercury column became, proving that air pressure diminishes with altitude.

Weighing air
It was Pascal's brother-in-law who undertook the altitude experiment on Pascal's behalf, demonstrating the existence of air pressure.

The grenade c1650

Glass grenades
Produced in the USA in the 1870s, these bottle grenades were filled with highly flammable chemicals.

Grenades would not have existed if the Chinese had not invented gunpowder in the 11th century. The first grenades were filled with black gunpowder, with a fuse extending at one end. The name came from the fact that they resembled the fruit of the pomegranate tree, known as a *grenade* in French.

The grenade became a popular weapon after 1650, when most major European armies created specialist corps of grenadiers. Despite the difficulty of lighting early models and their limited range as a hand-thrown weapon, grenades were soon in regular use on battlefields against close-range enemy positions. At Culloden in 1746, for example, after each sortie the Scots Highlanders were forced back behind low stone walls, where they came under musket and grenade attack from the English forces.

Trench warfare
An American GI throws a long-handled grenade at a German position during World War II.

Growing sophistication

Improvised grenades played a significant part in the Crimean War in the mid 19th century. They were made by filling glass bottles with gunpowder and nails, with a piece of string attached as a fuse. More effective devices followed, among them the British Mills bomb (see box above) and the German *Stielhhandgranate*. In use from the end of World War I to the end of World War II in 1945, this German weapon became known to British troops as the stick grenade or potato-masher because of its long throwing handle.

Grenades are still in use today, but hand-thrown models have now been supplemented by rifle and rocket-propelled versions. Among the most recent developments is the advent of purpose-built multiple launchers, capable of firing six grenades in rapid succession.

The post box
c1650

France took the lead in installing post boxes. A private mail service was established in Paris in 1603, allowing people to send letters to addresses across the country – but ironically the service did not apply within the city itself, so the only way of getting messages delivered in the capital was by messenger. In 1653 a state counsellor named Jean-Jacques Renouard de Villayer had the idea of creating a service that he called *La Petite Poste* (the 'Little Post'), consisting of wall-mounted boxes in the main thoroughfares of the city from which mail was regularly collected and delivered. Every letter had to bear a label indicating that carriage had been prepaid. The experiment was short-lived, but the idea of post boxes was subsequently taken up by the National Post; by 1692 there were six post boxes in Paris and twice that number by 1740.

Meanwhile, the idea behind *La Petite Poste* had crossed the Channel. In 1680 an entrepreneur named William Dockwra set up Britain's first penny post. Deliveries were made anywhere in London or, for an extra penny, within a 10-mile radius of the city.

Prussian blue
This late-19th-century post box from Prussia had display windows and revolving panels to show if there was any mail inside and when the next collection was due.

THE PILLAR BOX

The free-standing pillar box originated in Belgium, the first one appearing there in 1848. The novelist Anthony Trollope, a Post Office surveyor at the time, imported the idea to Britain and the first examples were trialled in the Channel Islands in 1852. Following their success there, pillar boxes were on the streets of the mainland a year later. The first cylindrical design was introduced in 1859. Hexagonal Penfolds were produced from 1866 to 1879. From 1874 all new boxes were painted red to make them more visible.

The vacuum pump 1650

One day in 1650 the physicist Otto von Guericke, mayor of the German city of Magdeburg, placed two bell-shaped metal receptacles rim to rim, then pumped the air out of them with a special pump he had devised for the purpose (consisting of a piston, a cylinder and a one-way valve). Two teams of eight horses were then hitched to each end and tried to pull the receptacles apart. They strained to no avail: the two hollow bronze hemispheres remained inseparably clamped together. When von Guericke reintroduced air into the vacuum he had created, the two halves separated easily.

Following on from the work of Galileo and Torricelli, von Guericke was seeking to demonstrate the force of atmospheric pressure. A few years later the English chemist Robert Boyle improved on von Guericke's design, subsequently claiming paternity of the pump that is used to this day to create vacuums (for example in light bulbs and cathode ray tubes).

Putting von Guericke to the test
In this 1672 engraving, two scientists prepare to test out von Guericke's experiment themselves using a barrel.

Money turns to paper

A simple piece of paper with a sum of money printed on it, anonymous and undated, the banknote gradually came to replace metallic coins as a means of payment. Long known in China, it was reinvented for Europe by a Dutchman living in Sweden, Johan Palmstruch. His Stockholms Banco was the first European institution to issue paper currency.

Chinese currency
Printed in the 14th century, this Chinese note (right) was worth 1,000 metal coins.

The *kopparplätmynt* coin and paper daler
Uniquely large and cumbersome, this square Swedish copper coin (above) was worth 10 silver dalers. If the 100-daler note (above right), dating from 1666, was cashed into kopparplätmynt *the pile of coins would weigh nearly 200kg.*

There is something reassuring about carrying coin of the realm. Anyone with gold or silver pieces in their pocket can be sure they have spending power and will not see the value of their coins crumble away to nothing. A coin possesses, in effect, an intrinsic value based on that of the metal it contains. So why opt to use notes instead? How did we come to put such confidence in a printed scrap of paper whose monetary value depends purely on the figure printed upon it?

Up to the middle of the 17th century, metal stocks were sufficient to meet all Europe's currency needs. But in a period of economic crisis, Sweden suddenly found itself in difficulties. The country had been embroiled in fighting wars and the extravagant lifestyle of the ruler, Queen Christina, had left the nation financially drained. Money was disappearing from circulation. What is more, Sweden's copper currency – the *kopparplätmynt*, in use since 1644 – was difficult to handle: some coin denominations measured 30 x 70cm and weighed a back-breaking 20kg.

The first banknotes

In this situation Johan Palmstruch, a Dutch merchant from Amsterdam, put in a proposal to establish a bank in Stockholm. King Charles X Gustaf, Christina's successor, gave his consent and the nation's first bank, the Stockholms Banco, duly opened its doors in 1657, protected by a troop of soldiers. Clients deposited their copper plaques and received paper deposit certificates in exchange. These bills were light and practical; henceforth money could change hands with a great deal less physical effort.

Three years later, the state reduced the weight of new *kopparplätmynt* plaques by 17 per cent. The effect of this was to devalue the paper bills against the metal that backed them, for the certificates were made out for the number of coins deposited, rather than for the metal's face value at the time. The bank's customers started clamouring to have their original, full-weight *kopparplätmynt* returned to them so they could sell them on for the copper they contained, which had in fact risen in value in the meantime. Fearful of running short of assets, Palmstruch was in 1661 granted the exclusive right to issue banknotes, known as *kreditivsedlar* ('credit paper'). Denominated in round numbers, these were convertible into copper but no longer had a specific exchange value in the metal. The first banknotes had come into use.

Marco Polo's story

To be more accurate, these were first banknotes in Europe – banknotes had already been in use in China for at least seven centuries. China was, after all, the place where both paper and printing were invented, so it is perhaps not surprising that it was also the first place to introduce paper money.

At the end of the Tang era (AD 618–907), merchants got into the habit of depositing their goods with the guilds they belonged to, accepting receipts in return. The system worked so well that the government decided to take up the idea. Tradespeople were henceforth encouraged to entrust their metal coinage to state-run establishments in exchange for notes. Under the Song Dynasty (960–1276) commerce was so prosperous in the province of Sichuan that there was a shortage of copper for minting coins. Some merchants began to circulate private currencies printed on mulberry bark. In 1024 the government imposed a state monopoly on the practice, and paper money gradually became the only legal tender. Under the Ming (1368–1644) the Ministry of Finance took responsibility for issuing banknotes, which thereafter bore a warning that anyone found guilty of forgery risked decapitation. But in its later decades the Ming regime was weakened by warfare and inflation, causing the public to lose confidence in the notes, which fell out of use.

Word of these developments in China had already reached Europe long before, carried there by the Venetian traveller Marco Polo. He described at length how the notes were made,

COMBATING COUNTERFEITERS

The first banknotes were rudimentary in design and easy to forge. To discourage fraud, governments decreed increasingly draconian punishments – ranging from deportation and amputation of hands to hanging and even boiling alive – for those found guilty of counterfeiting. In some countries the penalties were spelled out on the notes themselves. Yet all such moves proved futile; forgeries not only became more common but also more technically proficient. In 1695 the Bank of England introduced the first printed banknote to have no handwriting on it. In its wake 300 people were hanged for counterfeiting in London in 1697 alone, and that same year the Bank introduced the watermark. Banknotes continued to be at least partially handwritten in many countries; typically, as a proof of authenticity, the face value, the name of the recipient, the serial number and the cashier's signature would all be inked in. In later years banks relied on sophisticated technology to make printed currency as hard as possible to copy. Even so, notes continued to have a limited lifespan, because the best way of dealing with fakes was often to issue a new design.

One-guinea note *Issued by the Royal Bank of Scotland in 1777, the guinea note was the first to be printed in colour, adding another level of difficulty for forgers to overcome.*

even joking that while European alchemists had long sought unsuccessfully to transform base metals into gold, Chinese emperors had managed to turn paper into money. Polo was impressed by what he had seen, but at the time no-one in the West believed him.

Palmstruch's downfall

In the West bills of exchange supplied to merchants by bankers had been in circulation since the 15th century, and bankers' drafts were introduced in the 16th. But both these instruments depended on physical supplies of precious metals to back them up and did not involve any credit arrangement. Consequently, they did not increase the amount of money in circulation. That all changed with Johan Palmstruch's Swedish notes. The *kreditivsedlars* served to finance borrowings, yet their value relied entirely on the degree of confidence placed in the bank that issued them. If confidence wavered, clients might end up besieging the bank's branches to demand hard currency and bring the whole institution down. In the event, that was exactly what happened to Palmstruch. He issued notes to a value vastly superior to his cash assets, and in 1663 a wave of financial panic forced the Stockholms Banco to close its doors. Palmstruch was held responsible for the bank's failure and condemned to death. Reprieved, he died in prison in 1671, aged 60.

Imposing state control

The Bank of England – created in 1694 to help finance William III's war against Louis XIV of France – was the first to issue banknotes on a permanent basis. They became known as promissory notes as a result of the wording on them, promising to pay the bearer a specified sum on demand. The Bank of Scotland was established one year later.

Both institutions were privately owned, for at that point any bank could issue notes and for a time the number doing so in Britain and elsewhere grew rapidly, stimulated

Promissory note
A French note issued in 1720 (above) promises to pay the bearer 10 livres tournois. One livre tournois – so called because the currency was originally minted at Tours on the River Loire – was worth 0.31g of gold.

Seeking credit
A painting (left) shows the merchant-banker Jean-François Villiez (1722–74) receiving emissaries from Emperor Joseph II of Austria, who had come to his offices in the French town of Nancy seeking a loan. Rulers had a long history of borrowing money from private financiers in order to pay for major public works, government expenses and, of course, wars.

FINANCIAL PANIC IN FRANCE

It was a Scotsman, John Law, who introduced paper money to France in 1718, after Louis XIV's long wars had brought the country to the verge of bankruptcy. Law was a fugitive from his native land, having killed a man in a duel. In France he was authorised to found a bank to issue notes that were valid tender for public transactions, convertible at their original rate on demand. But Law printed too much paper money, causing a financial collapse that left most holders of the currency disastrously out of pocket.

In the early years of the French Revolution, some 70 years later, the revolutionary government took to issuing notes known as assignats, whose value was supposedly guaranteed against the worth of the goods confiscated from the aristocracy and clergy. The new regime needed funds to wage war against the republic's enemies, but they repeated Law's mistake, and the resulting inflation killed off the currency altogether.

As had happened earlier in other parts of Europe, these experiences made people distrustful of paper money in general. Banknotes only became generally accepted in France following the foundation of the Bank of France by Napoleon in 1800.

Shaky currency
A contemporary illustration (top) shows queues forming in John Law's bank in Paris to cash in their notes. The pile of notes and assignats (above) were issued during the Revolution.

The homunculus
Early users of microscopes sometimes let their fancy play tricks on them. Studying spermatozoa under the lens, the Dutch biologist Nicolas Hartsoeker announced in 1694 that he had found homunculi (right) – miniature human beings contained in the heads of male sperm.

The first great discoveries

Robert Hooke took an important step forward by recommending in the *Micrographia* that side lighting should be employed when examining objects against a dark background. His miscroscope was relatively modern in its design. Employing three separate glass lenses and with a slide for specimens, a focusing mechanism and a lens to concentrate the light, it was the first instrument of its type to have real powers of magnification of up to 30x. It was a composite creation, very different from earlier, simple microscopes, and Hooke's detailed description of the instrument served as a reference point at a time when the optical industry had not yet come into existence. Hooke rapidly grasped the possibilities that microscopes opened up, seeing them as essential tools for the study of living organisms.

In the comments that he made on the objects he examined, Hooke drew particular attention to the multitude of small 'chambers' or 'cells' that he observed in cork sections. This would be a crucial step in the biological sciences, since it marked the first use of the word 'cell', a fundamental building-block of all life forms. At the time Hooke did not recognise the full significance of his discovery. He thought that only plants had cells, and that their function was to transport liquids. Even so, he had opened the way for a new understanding of the natural world, and many scholars hastened to follow in his footsteps.

A Dutch scientist named Jan Swammerdam was studying reproduction when he realised that insects were not spontaneously generated, as had previously been thought. He demonstrated that they developed, like higher animals, through a progressive differentiation of the organs. His fellow-countryman Anton van Leeuwenhoek was the first to observe the world of protozoa and bacteria. The Italian Marcello Malpighi, the father of microscopic anatomy, discovered the capillary vessels through his observations of the lung, and was also the first to describe the tastebuds and the structures of the brain, skin, kidneys and glands, opening up the fields of human anatomy and embryology. All of these breakthroughs cast fresh light on the principal functions of the body – on respiration, reproduction, the circulation of the blood – and finally put paid to the old, imagined explanations of the body's workings drawn from Classical sources.

Continued improvements

In the 18th and 19th centuries great strides were taken in microscope technology. The problem of chromatic aberration was finally solved. Magnification, focus and resolution were all improved, as was the preparation of samples. As a result fresh discoveries were made – the nuclei of cells, cell division and chromosomes among them. For the first time

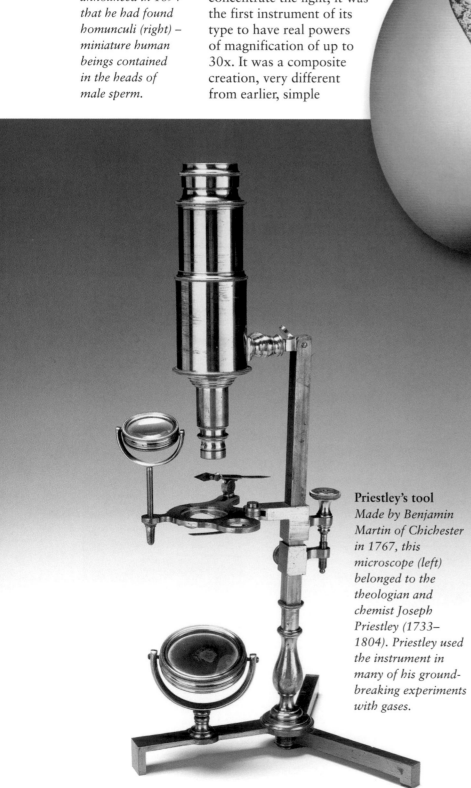

Priestley's tool
Made by Benjamin Martin of Chichester in 1767, this microscope (left) belonged to the theologian and chemist Joseph Priestley (1733–1804). Priestley used the instrument in many of his ground-breaking experiments with gases.

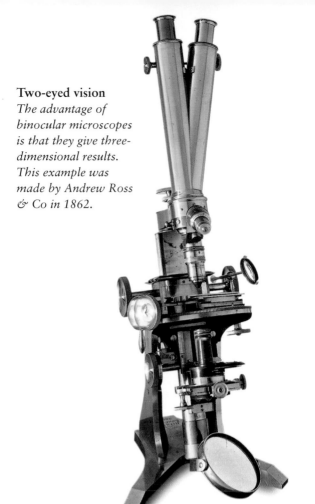

Two-eyed vision
The advantage of binocular microscopes is that they give three-dimensional results. This example was made by Andrew Ross & Co in 1862.

scientists could observe pathogens such as the bacilli responsible for leprosy and tuberculosis. Louis Pasteur used microscopes to study the process of fermentation. Medicine in particular benefited from the breakthroughs, but geology and chemistry were not far behind. Cellular biology, genetics, microbiology, bacteriology, crystallography – none of these disciplines would have existed without microscopes.

By the start of the 20th century optical microscopes were in widespread use. Magnifications ranging from 20x to 400x were common, with the best instruments reaching 1,500x. Major breakthroughs continued to be made, revealing previously undetected structures. Then, in 1931, a new and very differently conceived instrument took over the task of ultra-close observation. This was the electron microscope, in which electrons took the place of light, opening up a whole new dimension of discovery and knowledge.

IMPROVING THE IMAGE

As microscopy developed, scientists turned their attention to the quality of the image. In 1735 the English lawyer Chester Moore Hall solved the problem of chromatic aberration by using a combination of concave and convex lenses made, respectively, from crown and flint glass. In 1870 the German physicist Ernst Abbe, working for Karl Zeiss, demonstrated the effect that the type of light can have on contrast and resolution, as well as on the production of artefacts as a result of diffraction. In the 20th century the greatest progress was made in the focusing of ultraviolet microscopes (which offer a finer resolution) and of microscopy using a polarised light source, which reveals details that would be invisible in natural light. Other types of optical microscope, including phase contrast models and those relying on fluorescence, similarly play on the properties of light.

Revealing every detail
The modern technique known as scanning electron microscopy, which produced this ultra-close image of a spider's head (left), involves scanning the surface of the sample with a high-energy beam of electrons that interact with the atoms within the sample, producing signals that outline its topography.

The self-taught microscopist

By no means all the pioneers of the 17th century's scientific revolution had the benefit of a university education. A fine eye for detail, great manual dexterity, tireless curiosity and unusual longevity were the qualities that made the Dutch draper Anton van Leeuwenhoek the father of microbiology and one of the greatest scientists of his day.

In his workshop at the sign of the Golden Head, a young man was examining some clothing material through a small glass lens. The year was 1654 and the 22-year-old Anton van Leeuwenhoek had just set himself up in a drapery business in his native city of Delft. In the course of his apprenticeship in Amsterdam he had come across a device known at the time as a 'thread-counter', a magnifying glass set on a small stand. He had been astonished and intrigued to see how different things looked through the small lens. The back of his hand, for example, was transformed into a maze of fissures and furrows. The budding merchant liked to amuse himself by using the instrument to examine a host of different objects, without ever for a moment imagining the long-term consequences of his pastime.

An unexceptional man

Anton van Leeuwenhoek was born the son of a basketmaker who had an establishment close to the Leeuwenpoort ('Lion's Gate') in Delft. His father died in 1638, when Anton was barely six years old. Partly as a result, the boy received at best a limited education. But he was endowed with insatiable curiosity and a keen sense of observation, and in time these qualities would help him to become the leading microscopist of his day – the father of experimental microbiology.

Not much is known about van Leeuwenhoek's early years. He seems to have given up the drapery business at a relatively young age to try his hand at various municipal jobs that provided for his immediate needs. At different times he was an official in the Delft courts, a bailiff and a gauger, measuring the amounts of wine and beer sold by local publicans. In the mid 1660s his first wife, Barbara, died, and his grief at her passing may have been a factor in encouraging him to devote himself to science. At the time, Holland was in a golden age, buoyed up by its leading position in maritime commerce. In a continent swept by religious persecution, Holland's climate of tolerance attracted scholars and artists. Delft was flourishing as a centre of the sciences, painting and artistic craftsmanship.

Pioneers of microscopy
Van Leeuwenhoek (above, far left), like his pioneering contemporaries, made his own instruments to observe the world of the infinitesimally small. Robert Hooke, who was three years his junior, was the first person to use the word 'cell' (magnified above).

Making microscopes
Van Leeuwenhoek developed hundreds of microscopes in the course of his research. A replica of one of his very early tools is shown here (top centre), between two other experimental instruments made by the pioneering microscopist.

Birth of a vocation

In 1668 van Leeuwenhoek set off on the only substantial journey of his life. He travelled to London, where it seems likely that he had an opportunity to study Robert Hooke's famous *Micrographia*, which had been published three years earlier. Hooke's work described how to make a simple microscope and van Leeuwenhoek went on to apply the principle to produce hundreds of different microscope models, all employing a single biconvex lens inserted into a strip of metal and equipped with a needle attached to the stand that served to hold the samples. Over the years he elaborated this simple arrangement to the pitch of perfection. His most powerful instruments had a magnification of 500x and a resolution of a thousandth of a millimetre.

In the spring of 1673, the Dutch anatomist Régnier de Graaf sent a letter he had received from van Leeuwenhoek to Henry Oldenburg, the secretary of the Royal Society in London. Undeterred by the naval war that was then being fought between the two nations, de Graaf wanted to draw the attention of the international scientific community to the work of his compatriot, explaining in a covering note that 'one of our most ingenious minds, van Leeuwenhoek has managed to produce microscopes much better than any known so far'. Oldenburg read with fascination van Leeuwenhoek's descriptions of a mould 'resembling a sort of branch bearing leaves' and of a louse 'with a sting sticking out of its

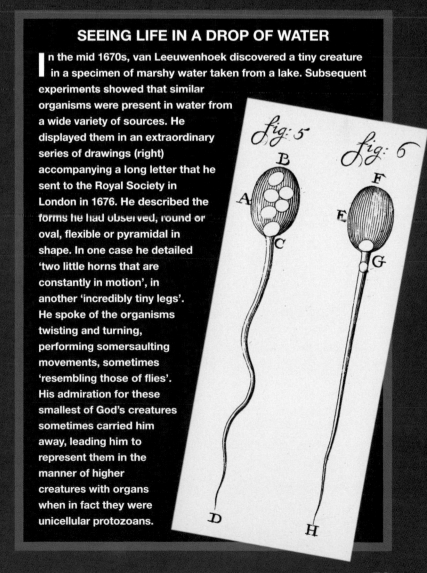

SEEING LIFE IN A DROP OF WATER

In the mid 1670s, van Leeuwenhoek discovered a tiny creature in a specimen of marshy water taken from a lake. Subsequent experiments showed that similar organisms were present in water from a wide variety of sources. He displayed them in an extraordinary series of drawings (right) accompanying a long letter that he sent to the Royal Society in London in 1676. He described the forms he had observed, round or oval, flexible or pyramidal in shape. In one case he detailed 'two little horns that are constantly in motion', in another 'incredibly tiny legs'. He spoke of the organisms twisting and turning, performing somersaulting movements, sometimes 'resembling those of flies'. His admiration for these smallest of God's creatures sometimes carried him away, leading him to represent them in the manner of higher creatures with organs when in fact they were unicellular protozoans.

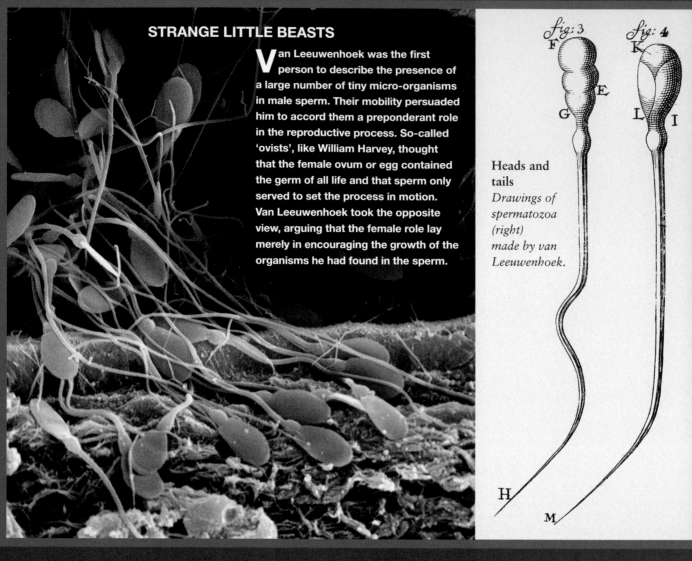

STRANGE LITTLE BEASTS

Van Leeuwenhoek was the first person to describe the presence of a large number of tiny micro-organisms in male sperm. Their mobility persuaded him to accord them a preponderant role in the reproductive process. So-called 'ovists', like William Harvey, thought that the female ovum or egg contained the germ of all life and that sperm only served to set the process in motion. Van Leeuwenhoek took the opposite view, arguing that the female role lay merely in encouraging the growth of the organisms he had found in the sperm.

Heads and tails
Drawings of spermatozoa (right) made by van Leeuwenhoek.

Simple solutions
Like the early instruments shown on the previous page, this second microscope designed by van Leeuwenhoek employed a single lens, here shown attached to a stand supporting a test tube which held the sample.

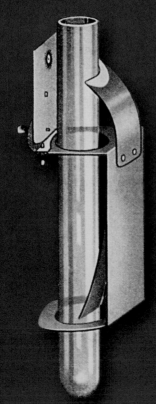

pointed nose'. Oldenburg subsequently published a translation of the letter in the Society's journal, *Philosophical Transactions*, and also wrote to its author encouraging him to keep on with his work.

Exploring a miniature world

Over the next 50 years van Leeuwenhoek sent almost 200 more letters to the Society. This correspondence, plus the letters to illustrious individuals such as Leibniz and Huygens that followed from it, form the totality of his published work. In them he gave a systematic description of all that he saw through the microscope. There seemed no limit to his curiosity or to the range of objects he chose to examine – drops of water, dental plaque, bodily fluids, insects, hairs, feathers, spices, gunpowder, grains, fruit. And thanks to the quality of the instruments he made, he kept on making fresh discoveries, among them the red cells in blood (1673), spermatozoa (1677) and yeasts in beer (1680). In 1689 he demonstrated the capillary circulation of the blood, first observed by the Italian Marcello Malpighi 28 years before. Van Leeuwenhoek is best known today for his precise descriptions of microscopic creatures – of rotifers, single-celled protozoa and bacteria – which opened the way for the new discipline of microbiology.

The rocky path to glory

In his long career van Leeuwenhoek had to overcome many handicaps. He had no academic education and spoke no Latin, nor indeed any language other than Dutch. He was a mediocre draughtsman, incapable of producing drawings to do justice to his observations. He was also secretive, refusing to reveal the details of his experiments or to describe exactly how he made his microscopes. As a result, some scholars regarded his work with scepticism. He even attracted bitter enemies, perhaps none worse than his Dutch rival and one-time pupil Nicolas Hartsoeker, who dismissed van Leeuwenhoek's talent as 'worse than mediocre'.

Yet van Leeuwenhoek was nothing if not determined. As proof that he had not made up his observations, he attached to his letters testimonials from reliable witnesses, including a clergyman, a jurist and even an archer, whose acuity of vision was judged to be beyond question. In some ways, his lack of education was an asset, for he started out with no preconceived ideas that might colour his observations. He won the respect of Robert Hooke, and in 1680 received the honour of being made a fellow of the Royal Society. Supported by his second wife Cornelia Swalmius, whom he had married in 1671, he built a reputation as a magician of the microscope and eventually became famous.

Throughout his long life he received visits from a succession of illustrious scholars, philosophers, churchmen, statesmen, even rulers, notably Peter the Great of Russia and Frederick I of Prussia. He died on 16 August, 1723, 'at the age of 90 years, 10 months and 2 days', as his inscription in the Oude Kerk of Delft specifies, with a degree of precision that van Leeuwenhoek himself would no doubt have appreciated.

AN UNRIVALLED OBSERVER

In 1981 the science writer Brian J Ford tracked down nine groups of original van Leeuwenhoek specimens that had languished unnoticed in the archives of the Royal Society. Having subjected them to extensive examination, using both ancient and modern instruments (including an electron microscope), he concluded not only that the samples were well prepared but also that van Leeuwenhoek revealed almost as much detail as modern equivalents.

The face of van Leeuwenhoek?
Johannes Vermeer painted The Geographer *(below) in 1669, possibly with his friend van Leeuwenhoek as his model.*

FAMOUS FRIENDS

Famed in their different fields for their powers of close observation, Anton van Leeuwenhoek and the painter Johannes Vermeer were both born in Delft in the year 1632, and are thought to have been friends. It has even been suggested that the microscopist may have served as the model for Vermeer's paintings of *The Astronomer* (now in the Louvre) and *The Geographer* (above), painted in 1668 and 1669 respectively. Vermeer was not blessed with the same longevity as van Leeuwenhoek, who acted as executor of Vermeer's will following the painter's death in 1675.

The spirit level c1660

Plumb level
If the air bubble settles between the two vertical rings, the surface is horizontal.

Ever since people first started constructing buildings, roads and canals, they have needed to be able to find plumb level. For a long time they did so by pouring water into various instruments and checking its flat surface. This worked fairly well for much of the time, but was always inconvenient and totally impractical in subzero temperatures.

Sometime around 1660 a French inventor named Melchisédech Thévenot had the idea of fixing a ruler within a sealed glass tube containing coloured alcohol – usually ethanol, as it does not freeze – together with a little air. When the tube-cum-ruler was placed on a level surface, the bubble of air positioned itself right in the centre of the tube. If the bubble was off-centre, the surface was not plumb. Today the support is normally made of aluminium, but the principle remains the same, and is as useful for professional builders as it is for weekend DIY enthusiasts.

Street lighting 1667

Venetian lamplighter
There was nothing new about street lighting in Venice when this drawing was made in the 18th century. The first experiments with illuminating the city's streets dated back to 1128.

The first organised experiments with municipal street lighting were made in London, Paris and Amsterdam during the 17th century, using candle-powered lanterns. In 1667 Gabriel Nicolas de La Reynie, then the police commissioner of Paris, attempted to create an effective lighting system in the capital when he had more than 5,000 lamps placed in the streets, each one emblazoned with the royal crest. At the time street lighting was considered to be first and foremost a security measure; it made the city safer and discouraged crime. In 1697 Louis XIV extended the arrangement to all the principal French cities. At first the local householders picked up the tab, but from 1704 the costs were met by the state. In the second half of the 18th century lights supplied with retroflectors were introduced, gradually replacing earlier models. Gas lighting was first employed in Europe – in London's Pall Mall – in 1807 and subsequently spread around the rest of the Continent; it had already been known in China for at least eight centuries.

LIGHTING LONDON

Several attempts to light the streets of London had been made before 1667, although not in a standardised fashion. As early as 1417 Sir Henry Barton, then mayor of the city, issued instructions for 'lanterns with lights to be hanged out on the winter evenings from Hallowtide to Candlemasse' (1 November to 2 February). Thereafter, householders were generally expected to provide some sort of illumination outside their houses on winter evenings, but the lamps were usually put out when the residents who lit them went to bed.

The Roberval balance

1669

A practical balance
Small models like this 19th-century Roberval balance (right) were designed mainly for kitchen use. The brass weights (below) ranged from 1g to 500g.

In 1669 a physician and mathematician named Gilles Personne de Roberval presented a 'new type of balance' to the French Royal Academy of Sciences. The device was an immediate success. In contrast to the hanging scales that had previously been used, the Roberval balance placed the measuring pans above the beam that linked them, giving the device a whole new degree of stability.

Shopkeepers' favourites

The device that Roberval put on show was not yet the classic pair of scales that would become familiar in kitchens all around the world. More work had first to be done by a British engineer, John Medhurst, at the start of the 19th century. Medhurst's improved versions were described at the time as 'French scales', acknowledging Roberval's contribution, although ironically by the mid 19th century Medhurst's models and their successors were being sold in France as 'English scales'. Further improvements were made in 1845 by Joseph Béranger, an inventor from Lyon, opening the way for a whole new range of developments.

By the late 19th century the Roberval balance in its updated forms had become standard equipment for shopkeepers, who liked it for its robustness, stability and precision. It was also popular with travelling salesmen, as it was easier to carry than most available alternatives. It remained the most common weighing apparatus until the 1980s, when it was finally ousted by digital scales.

STAYING LEVEL

The Roberval balance works on a principle of statics that remained unexplained for many years after its invention. When weights are placed in the pans, these rise or fall but they always remain level. The reason for this lies in the construction of the balance: two horizontal beams are joined by three vertical supports, the two outermost of which support the pans. (In models like the one above, the lower counterbeam is concealed in the base of the balance.) Together, these various elements form an articulated parallelogram, a figure that possesses an unusual degree of stability when the props move up or down. It was only in 1821 that the paradox was finally resolved by the French mathematician Louis Poinsot in the course of his work on the theory of couples.

Scales in the digital era
Most modern scales are of the spring variety (right), employing a strain gauge to measure electrical resistance.

SCALES AND BALANCES

Although both are used to weigh things, the principle behind the two types of device are different. Balances like Roberval's work by employing two pans on opposite ends of a horizontal lever. The object to be weighed goes into one pan, to be balanced against standard weights placed in the other. Spring scales, as the name suggests, exert pressure on a spring either by stretching or compressing it, to give a reading on a scale.

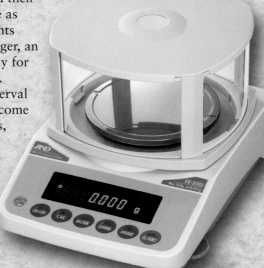

CHRISTIAAN HUYGENS 1629–95
The new Archimedes

A mathematician, astronomer and physicist, the Dutch scholar Christiaan Huygens had one of the finest minds of the 17th century, which he applied to the renaissance of scientific ideas that emerged during this period. He came on the scene at a crucial moment in the history of science and through his discoveries helped to bridge the gap between Galileo and Newton.

Portrait of an astronomer
A pastel drawing of Huygens made by Bernard Vaillant in 1666. Huygens was the first to recognise a ring around the planet Saturn, seen below in drawings he made at the time.

Diplomacy opens all doors, at least for those who know how to go through them. The career of the Dutch scholar Christiaan Huygens well illustrates the old adage. At the time of his birth in The Hague on April 14, 1629, everything pointed to his following in the footsteps of his grandfather, a secretary to William of Orange, and his father, Constantijn Huygens, a diplomat, philosopher and passionate lover of poetry who had been a friend of René Descartes. Respecting family tradition Christiaan, who could speak Latin fluently by the age of 9, went to the University of Leiden in 1645 to study law, then continued his studies at the Orange College in Breda. Fortunately for him and for astronomy, his father, who called him 'my Archimedes', realised that Christiaan was not cut out for a career in the civil service and endowed him with a private income, permitting him to devote himself to his true passion: science.

Discoveries in many fields

Having conducted research into falling bodies and written a treatise on squaring the circle, Huygens managed in 1655 to cast fresh light on a problem that had been troubling astronomers for almost half a century. From an observatory in Paris he trained a telescope of his own making on Saturn, and realised that its famous 'companions', observed earlier by Galileo, were not a couple of giant moons, as some had suggested, nor some mysterious flattened body as Galileo had thought, but rather a ring. This presented itself sometimes head on and sometimes in profile, depending on the planet's position in relation to the Sun. Twenty years later Giovanni Cassini would substitute 'rings' for Huygens' single ring, having observed gaps separating them.

In the same year Huygens discovered Titan, the largest of Saturn's moons, and described the rotation of Mars and the Orion Nebula. He was the first person to suggest that stars might be other suns, vastly distant from Earth and probably circled by planets of their own.

In the winter of 1656 Huygens invented the pendulum clock, providing a way of measuring time that was substantially more accurate than any known up to his day. In 1659 he turned his mind to the relationship between force and acceleration, and by associating the concept of free fall with rotational motion he came up with the concept of centrifugal force. Between 1661 and 1663 he worked, among other matters, on natural logarithms.

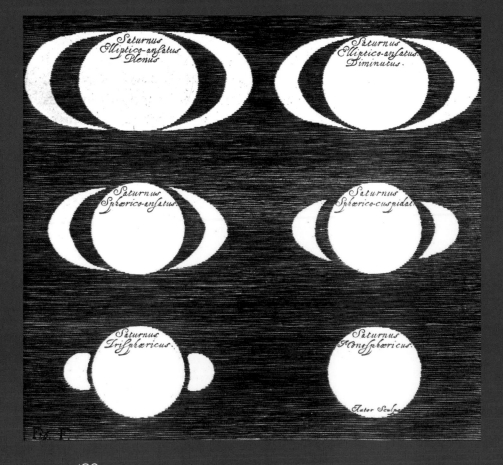

Achieving recognition

The newly-founded Royal Society in London made Huygens a member in 1663. Three years later Louis XIV's minister Jean-Baptiste Colbert invited him to Paris to set up the Royal Academy of Sciences, of which he was then appointed permanent secretary. The King granted him a pension, which gave him the financial freedom necessary to put his time in Paris (1666 to 1681) to good use. He defined inertia and elaborated theories of gravity and of light, thereby putting himself in conflict with Sir Isaac Newton. At the time, Newton and others adhered to the corpuscular theory that light was made up of tiny particles of insubstantial matter. In contrast, Huygens maintained that light was a wave. The issue would not be resolved in Huygens' favour until the 20th century.

Apart from a trip to London in 1689 to meet Newton – what the two men discussed went unrecorded, although they are known to have been admirers of each other's work – Huygens, whose health was deteriorating, spent the final years of his life in his native Holland. He died alone on July 8, 1695. More than three centuries later, this man of universal genius was accorded a most unusual honour: the module that travelled more than a billion kilometres from Earth to land on Titan in January 2005 bore his name.

AN HONOURED GUEST

Huygens was received in Paris with all due honour. He was awarded a substantial pension and was also the only academic to be offered lodgings in the Royal Library in the Louvre, where he was given an apartment with a laboratory attached. In gratitude, Huygens dedicated his 1673 treatise on pendulum clocks, *Horologium oscillatorium*, to Louis XIV.

Huygens' clock *These two views, face on and from the side, show the clock that was made to Huygens' specifications by Salomon Coster, a masterclockmaker in The Hague, in 1657.*

STRANGE MELANCHOLY

The illness that afflicted Huygens for many years remains something of a mystery, but it is known to have plunged him into periods of deep depression. Shortly before his death he described the symptoms in a poem:

The soul's languor, made one with the body's,
Infects the reason with a bitter melancholy
And leads whoever trusts it into folly.

Huygens' voyage to Saturn
Launched in 1997, the Cassini-Huygens space probe entered Saturn's orbit in 2004. The Huygens module separated from the orbiter to land on Titan, the largest of the planet's moons, sending back 350 photographs from the surface.

Training mirrors on the stars

Over the course of the 17th century, Galileo's heirs explored the cosmos with increasingly powerful telescopes, but the lenses available to them were of poor quality, producing blurred images. To remedy a defect that he wrongly considered irreparable, Isaac Newton took up an idea proposed by the mathematician James Gregory and replaced some of the lenses with mirrors.

When his invention was presented to the Royal Society in London on the morning of December 21, 1671, the young Isaac Newton was not present. In his place before the learned assembly stood his mentor, the mathematician Isaac Barrow, who explained the workings of the device. Despite the toy-like appearance of Newton's prototype, fashioned from wood and cardboard, the presentation was well received. On March 25 the following year, the Royal Society published *An Accompt of a New Catadrioptrical Telescope invented by Mr Newton*, which went on to become a source book for the modern science of optics. Newton's name was suddenly on everyone's lips.

The instrument under scrutiny was a tube, about 30cm long and 10cm in diameter, attached to a wooden sphere that moved freely in a socket, which in turn was fixed to a wooden base. At one end of the tube was a concave metal mirror 37mm across, with a focal length of 160mm; at the other end was the eyepiece with a small convex lens. The magnification was 38x – nothing extraordinary for its day – but the tube contained one more element that turned it into a 'reflecting' telescope. Newton's trademark as a telescope-maker was a small, flat, secondary mirror set between the main mirror and lens, which deflected light reflected by the main mirror at an angle of 45° toward the eyepiece.

Mounted by the master
A reflecting telescope built by Isaac Newton in 1672.

Newton's precursors

Astronomical observation had made huge strides since the start of the century. Galileo had popularised star-gazing. All across Europe aristocrats, artists and the educated middle classes had become intoxicated by the great celestial expanses, waxing lyrical on the infinite extent of the cosmos and the possibility of multiple worlds. Over a period of 60 years, ideas of the size of the universe had multiplied a hundredfold and conceptions of the human place within it had been transformed.

Newton was born in 1643, the year of Galileo's death. He grew up knowing that planets are other worlds, with their own mountains and clouds, and that some even have moons like the Earth. Then in 1655 Huygens had discovered Saturn's rings. Even so, Newton's invention of the reflecting telescope marked a breakthrough in the history of astronomy. The instrument itself was small and outwardly unremarkable, yet the Royal Society was right in claiming that it delivered remarkably clear images.

The fact was that ever since Galileo's day astronomers had been held back by the limitations of the available technology. Galileo's refracting telescope relied exclusively on lenses; a convex lens in the front part of the tube concentrated the light arriving from the stars, while at the back the eyepiece lens magnified the image for the observer. But the design suffered from a major weakness: chromatic aberration, or distortion of the resulting image, caused by the passage of the light through the glass. The light rays were refracted – deflected from their trajectory – at an angle that varied with their wavelength. The result was that multiple images of the same star formed at different distances for each colour, making the images fuzzy.

Astronomers responded to the problem by reducing the curvature of their lenses, which had the effect of disproportionately increasing focal length in relation to the lenses' diameter. The result was to lessen the aberrations and produce clearer images, but at a cost: the new telescopes became so long they were almost impossible to handle, requiring elaborate rigging systems that sometimes could only be moved with the help of several assistants. Huygens' telescope was 37m long; the refractor of the German astronomer Johannes Hevelius (1611-87) was 45m. Hevelius set his telescope in an observatory he constructed on the roofs of his three adjoining houses in Danzig. As well as his observations on the Moon and Sun he also discovered four comets. As a result he surmised that bodies such as these trace parabolic paths around the Sun.

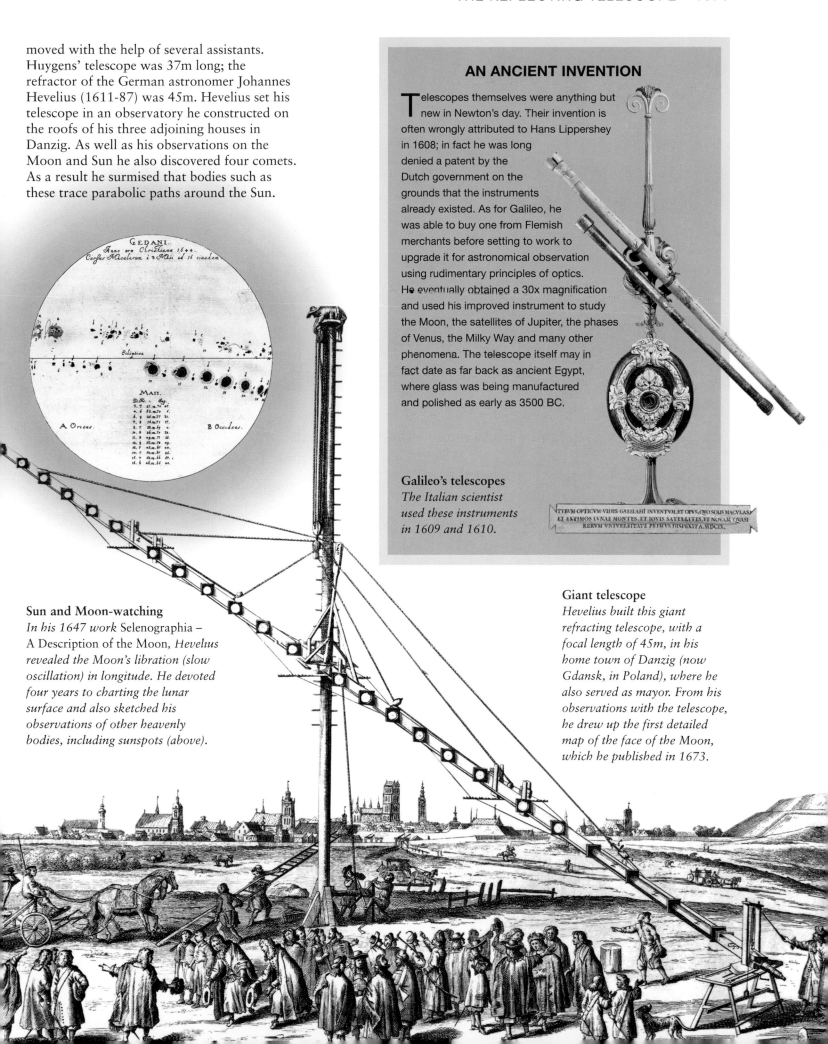

AN ANCIENT INVENTION

Telescopes themselves were anything but new in Newton's day. Their invention is often wrongly attributed to Hans Lippershey in 1608; in fact he was long denied a patent by the Dutch government on the grounds that the instruments already existed. As for Galileo, he was able to buy one from Flemish merchants before setting to work to upgrade it for astronomical observation using rudimentary principles of optics. He eventually obtained a 30x magnification and used his improved instrument to study the Moon, the satellites of Jupiter, the phases of Venus, the Milky Way and many other phenomena. The telescope itself may in fact date as far back as ancient Egypt, where glass was being manufactured and polished as early as 3500 BC.

Galileo's telescopes
The Italian scientist used these instruments in 1609 and 1610.

Sun and Moon-watching
In his 1647 work Selenographia – A Description of the Moon, *Hevelius revealed the Moon's libration (slow oscillation) in longitude. He devoted four years to charting the lunar surface and also sketched his observations of other heavenly bodies, including sunspots (above).*

Giant telescope
Hevelius built this giant refracting telescope, with a focal length of 45m, in his home town of Danzig (now Gdansk, in Poland), where he also served as mayor. From his observations with the telescope, he drew up the first detailed map of the face of the Moon, which he published in 1673.

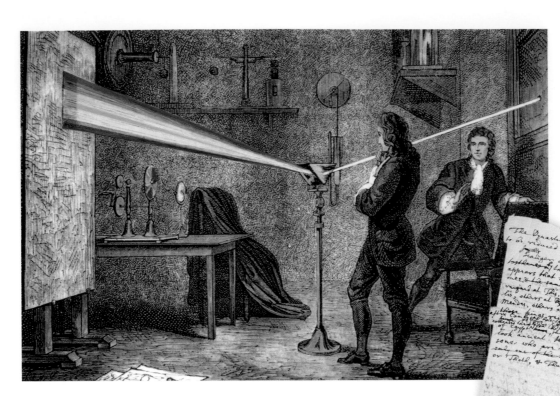

Optical experiment
*In 1666 Isaac Newton used
a prism to demonstrate the
dispersion of a beam of
white light into the colours
of the spectrum. He
recorded his observations
in 1704 in his* Opticks.
*A sheet from his notebook
is shown below.*

A fortunate error

Newton threw himself into the study of optics,
ingesting Descartes's writings on the subject
and carrying out experiments designed to help
him understand the passage of light through
lenses and prisms. He discovered that the
rainbow effect, produced when the Sun's rays
pass through a prism, stems from the light
rather than the device itself: in other words,
that white light is made up of all the colours of
the spectrum. But then he made an error. He
came to believe that the deflection of light and
dispersal of its colours through a transparent
body – the glass either of a prism or of a lens –
are correlated, independently of the object they
pass through. In fact, this is not so, but it
would take 80 years for students of optics to
realise it. The conclusion that Newton drew
was that there was no real prospect of
improving the traditional, refracting telescope.

This turned out to be a lucky mistake, for it
persuaded him to look for other solutions. He
turned his mind to mirrors, which reflect light,
like lenses, but do not refract it, thereby
avoiding chromatic aberration and creating a
clearer image. But if observation was easy
through the eyepiece of a refracting telescope,
there were seemingly insurmountable problems
to viewing an image through its reflecting
equivalent, for the mirror receiving the light
was where the observer's eye would normally
be, at the viewing end of the instrument.

James Gregory's contribution

Newton was not the first person to think of
using mirrors to correct chromatic aberration.

Descartes, the French physicist Laurent
Cassegrain and the Scottish mathematician
James Gregory had all played with the idea.
But their work remained in the realm of
theory, and the suggestions that they came
up with were beyond the capacity of the glassmakers of the day. One idea was to drill a hole
in the principal mirror and then install a
second facing mirror – either concave or
convex – so that the observer could view the
reflected light of the stars through an eyepiece
set in line with the hole. Gregory attempted to
build just such an instrument in 1663, but
soon ran up against the practical difficulty of
obtaining perfectly spherical polished surfaces.
It was left to Newton to bring the idea to
fruition by using an additional flat mirror to
deflect the reflected light into an eyepiece
placed on the side of the tube. The concept
was simple enough – and it worked.

Following the publication of details of
Newton's invention, a certain M de Bercé
wrote to inform the Royal Society that Laurent

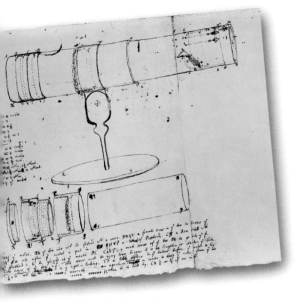

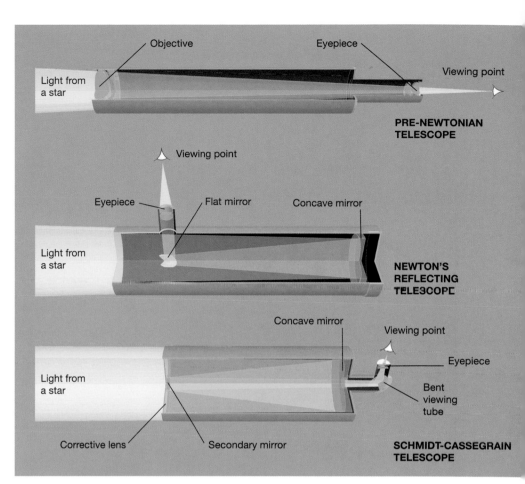

Objective — Eyepiece

Viewing point

Light from a star

PRE-NEWTONIAN TELESCOPE

Viewing point

Eyepiece — Flat mirror — Concave mirror

Light from a star

NEWTON'S REFLECTING TELESCOPE

Concave mirror

Viewing point

Eyepiece

Bent viewing tube

Light from a star

Corrective lens — Secondary mirror

SCHMIDT-CASSEGRAIN TELESCOPE

Newton's breakthrough

The drawing above illustrates the components of the reflecting telescope, also shown in the diagram (right). A pre-Newtonian instrument is illustrated above it and a Schmidt-Cassegrain model below.

Cassegrain had already outlined a design for a reflecting telescope more effective than Newton's, and that details had been duly published in the papers of the French Royal Academy of Sciences. Newton responded icily in the Royal Society's journal of May 1672, pointing out that James Gregory had proposed a model similar to Cassegrain's as early as 1663, but like Cassegrain had failed to produce a working model. Newton concluded his response by saying that Cassegrain's solution needed to be made operational to claim precedence over his own instrument.

VARIABLE STARS

Most stars, like our own Sun, have a relatively constant luminosity, but some vary in their brightness. Skywatchers noted this phenomenon even in antiquity but mostly chose to ignore it, as received opinion at the time held that the celestial sphere was perfect and unchanging. The question of variability was not seriously addressed until the 17th century, when the Danish astronomer Tycho Brahe observed a supernova. In the years that followed Brahe's breakthrough, improvements in telescope design revealed more and more variable stars. Today more than 40,000 have been catalogued, usually classified into two main types: those whose luminosity actually changes over time and those that merely seem to grow brighter or less bright, perhaps as a result of periodic eclipses.

Cassegrain's alternative

A portable version of the Cassegrain telescope (below) was built by Jesse Ramsden of London between 1762 and 1774.

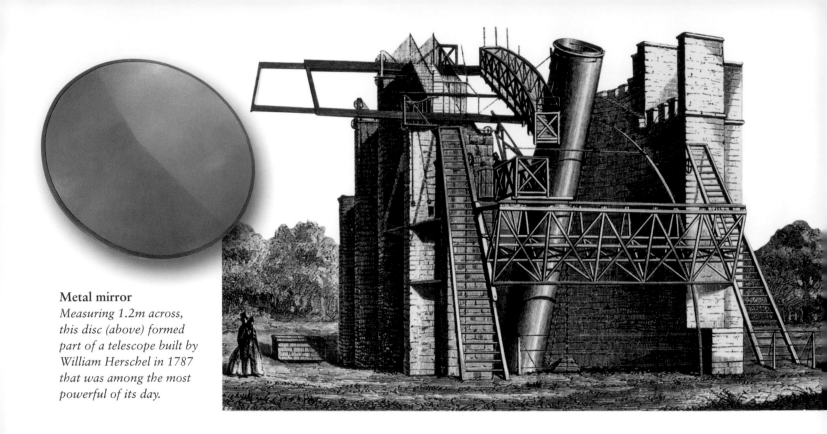

Metal mirror
Measuring 1.2m across, this disc (above) formed part of a telescope built by William Herschel in 1787 that was among the most powerful of its day.

The Parsonstown Leviathan
The gigantic telescope above was built for the Earl of Rosse in 1844 and installed on his estate at Birr Castle in Ireland. The world's largest telescope at the time, it featured a mirror weighing 3 tonnes and had a 6400x magnification.

Searching for meteorites
An American astronomer scans the skies for meteorites that could pose a threat of collision with the Earth (left).

The trouble with metal mirrors

Posterity has come to acknowledge the contibution of all three men and modern variants of their instruments are still in use around the world. Yet for two centuries after Newton's design was unveiled, it was more often praised than used. Once again, the problem lay in the technology of the day. Mirrors at the time were made not of glass but of metal alloys that quickly tarnished. To construct his instrument, Newton mixed six ounces of copper to two of tin and one of arsenic. Long hours of polishing produced the best reflective surface he could manage, but one that was still less good at transmitting light than lenses. In practical astronomy, Newton's invention was less effective than he hoped.

Meanwhile, progress was made on the earlier refracting telescopes. In 1758 the problem of chromatic aberration was finally solved by an English optician named John Dollond, whose name is still commemorated up and down the land in the high street optician, Dollond & Aitchison. Before being appointed optician to the King in 1761, Dollond had found a way of constructing achromatic lenses by combining two separate lenses, one convex and one concave, within a single objective. The two lenses were made from different types of glass with different refractive indices. Refracting telescopes using compound objectives of this type gained in quality and were easy to use.

Reflecting telescopes returned to favour in the late 19th century after French scientist Léon Foucault replaced the metal mirrors with

THE PROBLEM OF ASTRONOMICAL SEEING

As telescopes have got bigger, they have increased their capacity to gather in light, permitting their operators to see smaller and smaller objects. But the growing size of telescopes has not improved their capacity to reveal detail – their so-called 'angular resolution'. The reason lies in the Earth's atmosphere, which is constantly disturbed by the movements of masses of air, creating a turbulence that deflects light rays arriving from space. Astronomers use the word 'seeing' to address the problems of blurring, loss of detail and reduced contrast that the atmospheric disturbances can cause, making stars appear in their instruments not as fine points but rather as smears. In all these ways, atmospheric turbulence can prevent even the most advanced and powerful telescopes from attaining their optimal angular resolution. Ever since the 19th century, astronomers have sought to combat the effects of turbulence by building observatories at high altitudes, seeking thereby to minimise the distance that light from space has to travel through the atmosphere. All the biggest modern telescopes have been built on the summits of mountains or volcanoes, such as the observatories on Mauna Kea in Hawaii at an altitude of 4,200m and at La Palma in the Canary Islands (2,400m). Angular resolution, and therefore image quality, is two or three times better at these sites than at ground level.

Halley's comet
The comet's passage in 1986 was photographed from Mauna Kea on Hawaii (above), whose observatories house a dozen telescopes among the most powerful in the world.

Telescope in space
Placed in orbit by the American shuttle Discovery *in 1990, the Hubble Telescope (below) circles the Earth, sending ground-breaking images of the universe back to the planet's surface 600km below.*

ones of silver-coated glass, opening the way for the development of today's astronomical telescopes. Soon after, people stopped making refractors with apertures more than 1m in diameter, because the lenses were so heavy they tended to warp. Mirrors suffered no such constraints, thanks partly to their supports, and as a result telescopes became bigger and bigger. In 1909 the largest optical reflecting telescope in the world had a mirror diameter of 1.5m. By the end of the century the figure had risen to 10m and European astronomers are currently considering building an instrument with a 42m objective.

SPACE OBSERVATORIES

After the era of sky-watching with the naked eye and then the age of optics, the 20th century introduced a new epoch of observation from space. Telescopes were placed in orbit around the Earth, the best known of them being the Hubble. Beyond the reach of atmospheric disturbance, these instruments sent back images of unprecedented clarity. The new powers of observation they have opened up are providing evidence that may eventually enable astronomers to validate or disprove theories about the universe, among them the long-running debate concerning its age.

Binoculars 1671

Early binoculars
A prototype binocular telescope devised in 1671 by a Capuchin monk, Chérubin d'Orléans.

Galileo's microscope and Newton's telescope shared one thing in common: both had a single eyepiece, so the user studied the image cast by the lens or mirror through only one eye. It was a natural next step to link two optical tubes in such a way as to permit the observer to use both eyes, thereby viewing objects at a distance in a seemingly more natural manner. A monk by the name of Chérubin d'Orléans made a prototype pair in 1671, describing his invention six years later in a work entitled *Perfect Vision*.

Binocular design and quality was not significantly improved for nearly two centuries. Then in 1854 the Italian optician Ignazio Porro patented his image erecting system incorporating a double prism arranged in a Z shape. This allowed the physical length of the binoculars to be less than the focal length of the objective, so improving magnification and image quality. Further refinements followed, but the original Galilean design continued to be used – as it is to this day – for opera or theatre glasses.

A useful military tool

Military planners were quick to grasp one fact about the new invention: that binoculars, like the eyes of humans and other predators, provide vision in depth. The objectives in each tube receive light from the image at a slightly different angle to the other tube, causing the brain to automatically reconstitute natural perspective as it analyses the disparity. In the average person, the irises are set 6.5cm apart, but the gap between the objectives in binoculars can be widened, augmenting the perspective effect. During the First World War instrument-makers exploited this to create viewing instruments with lenses 1m apart, giving a hyper-realistic view of the battlefield.

Such innovations were of little use to astronomers. From an optical point of view, the stars are infinitely distant and so have no sense of depth to an observer on Earth. As a result, binocular telescopes have never been more than curiosities. They are difficult to make and to focus and are rarely used for studying the sky.

NIGHT VISION

Among the many advances made in optical technology over the years have been techniques enabling people to see in the dark. Night-vision binoculars utilise a variety of mechanisms – including photomultiplier tubes and thermal imagers – that are sensitive to a wide spectrum of light, from visible light to infrared. In practice, the user ends up viewing an enlarged electronic image projected onto a phosphorescent screen. The instruments are popular with the military and also with naturalists observing nocturnal animals.

Double-lensed periscope
A German soldier on campaign in the Libyan desert in 1942 peers through raised binocular sights that enable him to scan the surrounding terrain without breaking cover.

CASTING PLATE GLASS

In the 1680s a French glassmaker named Louis Lucas de Nehou devised a new process for producing plate glass by pouring the molten liquid onto special tables, where it was then rolled out flat. The resulting material was carefully polished with abrasive sands and felt-covered discs, then coated on one side with reflective metal. The technique was used to create the great mirrors that decorated the stately homes of the day, notably the Palace of Versailles outside Paris.

Lead glass 1675

Lead glass, also known as lead crystal, was born of a ban. In 1615 King James I (VI of Scotland) forbade the industrial use of firewood in order to combat deforestation, obliging glassmakers to turn to coal to fuel their furnaces. To avoid accidentally colouring the glass, they had to work in closed crucibles, which slowed down the melting process. A glassmaker named George Ravenscroft had the idea of adding lead oxide to create a clearer, more transparent product. He applied for a patent in 1675, and the crystal glass industry was born. Ravenscroft's successors improved on his formula by increasing the amount of lead oxide in the mix to as much as 30 per cent. The technology soon spread and the glassworks of Baccarat in France, founded in 1764, established a reputation for excellence.

Hall of Mirrors
The most famous room in the Palace of Versailles saw the signing of the treaty at the end of the First World War. It is 73m long and lit by 17 windows, whose light reflects from 357 mirrors.

Ice cream vendor
This ancestor of today's ice-cream drivers plied his trade in Italy in the 18th century. He kept his wares cool in ice buckets.

Ice cream 1686

The first known reference to ice cream in Europe dates from 1686, when it was served as a dessert to King James II. In the same year a Sicilian restaurateur named Procopio dei Coltelli opened the Procope, the first café to serve the delicacy in Paris. Wits and aristocrats were soon rubbing shoulders there, attracted by the 80 or so different flavours on offer, including rose, elderberry and grilled orange blossom.

Ice cream remained a luxury item until the late 19th century. The first cornets made their appearance in 1904 at the St Louis World's Fair in the USA. Twenty years later the US entrepreneur Christian Nelson invented the choc-ice bar.

AN ORIENTAL DELIGHT

Three thousand years ago the Chinese were already enjoying mixtures of rice and sugared milk kept chilled in buckets beneath a layer of snow and saltpetre. The Arabs later adapted the process to create *sharbat*, the ancestor of modern sorbets.

Explorers of the vegetable kingdom

Unlike earlier naturalists, John Ray, the father of British natural history, chose to study plants not for their medicinal use but along scientific lines. In later years, the study of botany became a vast discipline dedicated to classifying and understanding the entire living world.

The story of plants
John Ray's Historia Plantarum *described almost 18,000 plants. Illustrations of plants were already being made. This depiction of a herbaceous plant of the lily family (right) was done in the 16th century.*

In 1686 the English naturalist John Ray published the first instalment of his magisterial *Historia Plantarum* ('History of Plants'). The work, finally completed in 1704, just a year before his death, covered all plants known at the time. Publication of Ray's work marked a decisive step in the progress of botany, a word that had first come into use around the year 1610 to describe the study of the vegetable kingdom. By attempting to classify plants in a systematic manner, Ray turned the investigation of plants into a science.

Unlike most naturalists of his day, Ray was not a doctor, and so did not approach the study of plants from a medicinal angle. Instead, he was ordained as a clergyman, spending most of his life in the Essex village of Black Notley where he was born. A devout Christian, he was an emblematic figure in an age of transition, when a belief in the divine origin of the world only encouraged the desire for a fuller understanding of its phenomena. In Ray's case the urge expressed itself in a range of scientific enquiry as well as in many research trips, most notably a tour of Europe made in company with the ornithologist Francis Willoughby between 1663 and 1666. John Ray thereby took his place in a line of scholars who, from the start of the 16th century, had dedicated their lifework to the study of nature and the living world.

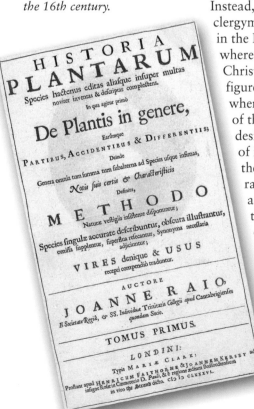

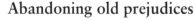

Abandoning old prejudices

For all their fierce curiosity, stimulated by the progress of humanism and by the great discoveries that had pushed back the limits of the known world, early botanical researchers still found it difficult to free themselves from the received body of ideas passed down from Classical times. They continued to promote a descriptive approach to their subject that was interlaced with myths and folklore. Encyclopaedists, such as the German-speakers Conrad Gesner, Leonard Fuchs and Hieronymus Bock, drew up compendia of knowledge based on existing sources, while also showing a new inclination to observe nature at first hand.

The first real attempt at scientific plant classification was made by Andrea Cesalpino, physician to Pope Clement VIII, whose treatise *De plantis* ('On Plants'), published in 1583, sought to relate each plant described to similar species rather than to potential uses. Following the example of Aristotle, Cesalpino aimed to

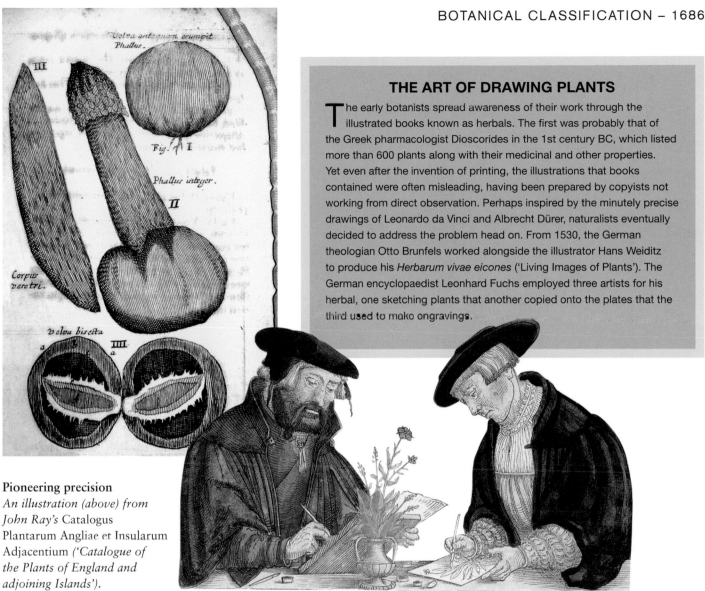

THE ART OF DRAWING PLANTS

The early botanists spread awareness of their work through the illustrated books known as herbals. The first was probably that of the Greek pharmacologist Dioscorides in the 1st century BC, which listed more than 600 plants along with their medicinal and other properties. Yet even after the invention of printing, the illustrations that books contained were often misleading, having been prepared by copyists not working from direct observation. Perhaps inspired by the minutely precise drawings of Leonardo da Vinci and Albrecht Dürer, naturalists eventually decided to address the problem head on. From 1530, the German theologian Otto Brunfels worked alongside the illustrator Hans Weiditz to produce his *Herbarum vivae eicones* ('Living Images of Plants'). The German encyclopaedist Leonhard Fuchs employed three artists for his herbal, one sketching plants that another copied onto the plates that the third used to make engravings.

Pioneering precision
An illustration (above) from John Ray's Catalogus Plantarum Angliae et Insularum Adjacentium *('Catalogue of the Plants of England and adjoining Islands').*

discover the essence of things, while sticking to the notion that they owed their nature to God's divine plan.

Seeking order in nature

John Ray took on board all the work of his great predecessors. He was well informed about the intellectual life of his time, and was intrigued by a hypothesis put forward by the German botanist Rudolf Camerarius, director of the botanical gardens at Tübingen, that plants reproduced sexually. From Camerarius's day it became generally accepted that plants were generated by members of their own species and could not be made fertile by different species. It followed from this that close observation of the organs of any given plant could enable researchers to determine the group or class to which it belonged. So by studying the structure of leaves, Ray was able to distinguish monocotyledons, such as grasses, from dicotyledons with two embryonic leaves within their seeds rather

Herbalists at work
Two illustrators at work on Leonard Fuch's De historia stirpium *('On the History of Plants'), a herbal published in 1542 that listed 487 species.*

Spotter's guide
The frontispiece of Eléments de botanique *('Elements of Botany'), a work by the French naturalist Joseph Pitton de Tournefort (1656–1708). This was the first work to clearly distinguish genera from species.*

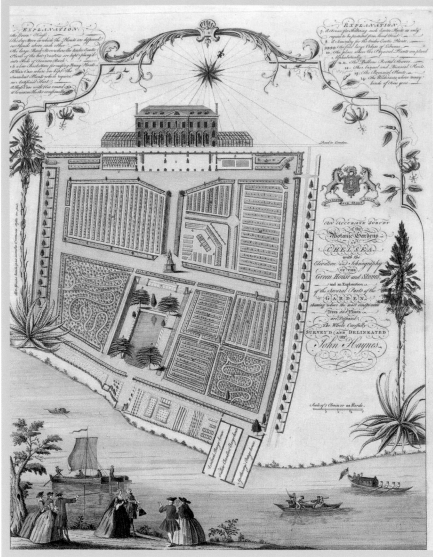

ALL NATURE IN A GARDEN

The first botanical gardens – such as the one created by Matthaeus Silvaticus in the early 14th century at Salerno in Italy – brought together medicinal plants for healing and teaching purposes, on the model of the herbal gardens that adjoined many monasteries at the time. By the late 16th century the growing interest in botany stimulated the foundation of a new group of gardens in Europe, this time inspired by the work of plant collectors and the desire to acclimatise non-native species brought back by travellers from distant lands. The first such garden to be established in England was the University of Oxford Botanic Garden, founded in 1621, which predated Paris's celebrated Jardin des Plantes by five years. It was set up with money donated by the 1st Earl of Danby to serve as a physic garden, designed to grow plants for medical research.

The Chelsea Physic Garden *(left) was founded in 1673 on the initiative of the Worshipful Society of Apothecaries.*

than just one. In 1694 the French naturalist Joseph Pitton de Tournefort would build on Ray's work in his *Eléments de botanique* ('Elements of Botany'), proposing a more ambitious scheme of classification that took into account the concepts of genera and species as well as plant varieties.

The Linnaean path

Thereafter botany never looked back. In the 18th century research became more specialised, focusing on reproduction, nutrition and respiration in the vegetable kingdom. Yet the desire to bring order to the study of nature in no way diminished. The Swede Carl Linnaeus was inspired to draw up a complete inventory of all known animal and plant species. His approach was criticised at the time by some botanists who would have preferred a more empirical approach to the classification of genera and families, but even so Linnaeus's work marked a crucial step forward by its sheer thoroughness. The Linnaean system of

Fungal flora
A page of different varieties of mushroom (right), from a work by the pioneering Swedish botanist Carl Linnaeus (1707–78).

nomenclature provided scholars with a common terminology in which each plant had a double Latin tag, the first word indicating its genus and the second its species.

Towards a theory of evolution

Linnaeus's work brought the first phase of botanical history to a close. It had seen simple observation of nature transformed into a scientific discipline. Botanists now benefited from an agreed taxonomy that was constantly enriched by the discovery of new species. Yet their world view was still shaped by the belief, drawn from the Bible, that creation was eternal and unchanging.

Towards the end of the 18th century a French naturalist, Jean-Baptiste Lamarck, began to contemplate the extinction of certain species, notably as a result of climate change. He eventually arrived at the notion of transmutation – that living things could somehow adapt to their environment, for instance by developing new organs or modifying existing ones. He cited the example of the

giraffe, whose neck had grown long so it could browse on the leaves of tall trees, the only food available to it on the dry African savannah where it lived. He also put forward the notion that acquired characteristics might be inherited, putting the idea of evolution on the table, to be fiercely contested by early creationists. Soon, Charles Darwin would build on these ideas, but reject Lamarck's view of the 'transmutation' of species in favour of his own concept of natural selection.

Native species *An illustration from the* Flore française *('French Flora') of Jean-Baptiste Lamarck showing fruits that 'grow naturally in France'.*

CLASSIFYING THE ANIMAL KINGDOM

The word 'zoology' came into use in the mid 18th century to designate the study of animals. In the Middle Ages most scholars relied on the sometimes fantastic descriptions contained in the *Natural History* of Pliny the Elder, written in the 1st century AD; for the most part they remained ignorant of Aristotle's more scientific attempt to classify creatures by their reproductive habits. The *Physiologus*, a 2nd-century bestiary, was also influential, despite the fact that it included mythical beasts like the griffin (a lion with the head and wings of an eagle) alongside real creatures. In the 16th

century the Swiss naturalist Conrad Gesner tried to bring order to the subject by drawing on all existing sources to compile a comprehensive taxonomy, giving each animal a Latin name consisting of a genus and a qualifying adjective. A century later Linnaeus applied a similar system to the entire animal and vegetable kingdoms. A French naturalist, the Comte de Buffon, considered Linnaeus's classification too systematic, so for his *Natural History* he devised an approach based on observation of the physiology and behaviour of different species, linked to the conditions in which they lived. Subsequent advances included the proposition, put forward in 1839 by Matthias Jakob Schleiden and Theodor Schwann of Germany, that cells are the basic structural units of all living things. Modern taxonomy, or cladistics, classified organisms still further according to their shared ancestry, focusing on their evolutionary relationships.

Marine creatures *from the* Historiae Animalium *of Conrad Gesner, a pioneering Swiss zoologist.*

A single force linking Earth and the heavens

At the start of the 17th century, Galileo proposed one set of rules for the motion of earthly bodies and Kepler another for the heavens. Newton unified the two in 1687, showing that the fall of an apple and the Moon's orbit both obey a single law – that of universal gravitation.

Two planets
Johannes Kepler made this drawing (right) of the conjunction of Jupiter and Saturn, which he observed in December 1603.

Pondering the Moon *Shown here in an aquatint of 1791, Bernard le Bovier de Fontenelle (1657–1757) was an author and Enlightenment thinker with a particular interest in astronomy. His Entretiens sur la pluralité des mondes ('Conversations on the Plurality of Worlds') popularised the heliocentric model of the universe put forward by Copernicus.*

People observing the night sky from the Earth in the 17th century saw an unconnected world. According to their way of thinking, nothing that occurred up there had any link to what was happening around them, for the Earth had one set of rules and the heavens another. When they beheld the Moon, they saw a suspended body following a uniform circular course at a constant speed, symbolising celestial harmony. In 17th-century eyes the motion of the Moon was not related in any way to the movement of bodies here on the terrestrial sphere, which either moved forward horizontally or fell vertically to Earth – or followed some combination of the two. There was no parallel to be found in earthly terms for the uniform circular motion of the Moon, just as there was no concept of up or down trajectories in the sky. Aristotle had said so some 2,000 years ago, and nobody felt inclined to challenge the great man's authority.

Separate rules for separate worlds

By mid century, individuals who had read the works of Galileo or Kepler might have had a slightly more sophisticated understanding. They would have recognised that both terrestrial and celestial bodies obeyed mathematical laws, which Galileo had spelled out for the earthly sphere and Kepler for the stars. Galileo had shown that all earthly bodies, whatever their weight or mass, fall at the same speed with uniform acceleration. He had also proposed that a body moving forward horizontally tends to remain in motion at a constant velocity, unless some external force such as friction or air pressure slows it down.

Meanwhile, Kepler had discovered that the motion of the planets is not uniformly circular, as the ancient Greeks had believed, but is elliptical and of varying speed, with periods of acceleration and deceleration. The two men both employed mathematical laws and principles to argue their case. But there was no link between the laws that they proposed. The upper and lower realms remained quite distinct and separate.

Phases of the Moon
Newton drew inspiration for his theory of universal gravitation from the Moon's movements and appearance (left).

Epoch-making theory

This situation changed once and for all in 1687 with the publication of Newton's *Principia Mathematica*. In the new theory of universal gravitation that he propounded, the laws governing Earth and sky became simply two facets of a single coin.

In Newton's view, gravity was a force engendered by mass that operated over distance to attract matter to matter. This force affected every solid body in the universe. It not only made apples fall to the ground but also pulled the Moon around the Earth and the Earth around the Sun. In the Newtonian scheme of things, gravity was constantly drawing the Moon towards the Earth just as it attracted an apple falling from a tree, the difference being that the Moon always missed its target. Gravitation was a universal law for the movement of masses – one that applied equally on Earth and in the celestial sphere.

Intellectual gravity
A 1763 cartoon shows Newton's concept of gravity being used to measure brain weight, with geniuses plunging to the ground, head down, and idiots floating upright.

Thinker at work
Artists often portrayed Isaac Newton in meditative pose under an apple tree, helping to promote the legend around the inspiration of his gravity theory. This one (below) was painted by Robert Hannah in 1905.

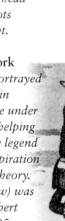

NEWTON'S APPLE – BIRTH OF A LEGEND

Newton himself spread the idea that a falling apple in an orchard had helped to inspire his theory of gravity. A friend of Newton's, the antiquarian William Stukeley, recalled in a memoir a conversation he had had with Newton shortly before the scientist's death. It seems Newton had mentioned the story, recounting how it led him to wonder why a falling apple always hit the ground: 'Why should it not go sideways or upwards, but constantly to the Earth's centre?' Another memoir-writer put a slightly different spin on the story, claiming that it led Newton to speculate whether the force that pulled the apple to Earth might not extend much further than usually thought: 'Why not as high as the Moon, said he to himself, and if so that must influence her motion and perhaps retain her in her orbit?'

THE LANGUAGE OF MATHEMATICS

Newton held that every body or mass – whether a cannonball or the entire planet Earth – engenders a force of attraction over distance. The forces of two separate bodies combine to create a reciprocal force that increases with the size of the bodies involved and diminishes with distance. His theory can be expressed mathematically: the force of attraction between two bodies is proportional to the product of their mass and inversely proportional to the square of the distance between them.

Forces within forces
Newton drew this sketch in the Principia Mathematica *of 1687 to illustrated the Cannonball Theory.*

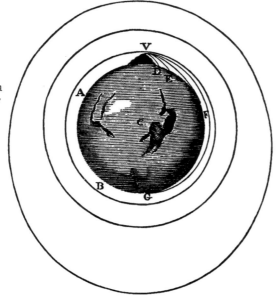

From Galileo's cannonball ...

Newton came up with the theory of gravitation by following up an idea originally proposed in 1604 by Galileo. The Italian scientist was trying to cast light on the motion of terrestrial bodies by imagining a cannon positioned on the summit of a hill in order to fire a cannonball horizontally. The projectile, Galileo asserted, would follow a trajectory that was the result of two different elements: a horizontal forward motion at constant speed and an accelerated downward motion.

... to Newton's satellite

Sixty-one years later, Newton extended Galileo's idea by imagining that the cannon had unlimited firepower. Each time the gun was fired the power was augmented to increase the speed of the cannonball's horizontal trajectory, causing it to travel further than the last ball. Despite the greater distance, Newton realised that the ball would nonetheless take the same time to touch the ground, because the vertical distance it had to fall was the same.

When the shots became very powerful, and the distances therefore very great, Newton conjectured that a new factor came into play: the curvature of the Earth. The cannon might fire the ball horizontally as before, but the Earth would fall away from under it in the course of its flight; in falling, the ball would itself follow the curve. With each more powerful shot, the ball would overfly a greater portion of the globe before eventually touching down. In time it would circle the entire planet before finally crashing down at the foot of the hill from which it was fired.

Newton's visualisations did not stop there: increase the power of the shot even more, he argued, and the ball would not just circle the Earth and return to its starting point, it would continue at the same speed it had started out, going on to make a second and perhaps many more revolutions. Newton's cannonball had become a satellite of Earth just like the Moon. In his view this was solely as a result of the force of gravity, which explained the ball's movements. The rules governing Earth and the heavens had become one and the same.

A PROBLEM FOR EINSTEIN

Newton himself was the first to point out the limits of his understanding of gravitational force. In 1687 he wrote: 'That one body may act upon another at a distance through a vacuum without the mediation of anything else ... is to me so great an absurdity that, I believe, no man who has in philosophic matters a competent faculty of thinking could ever fall into it.' But it was the best understanding available until Einstein extended Newton's ideas by proposing the theory of special relativity in 1905.

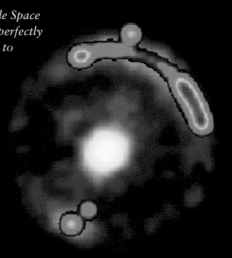

Gravitation between galaxies
This image taken by the Hubble Space Telescope shows two galaxies perfectly aligned with one another so as to look like one body. The green arc is created by deformation of the light emitted by the furthermost galaxy, caused by the gravitational pull of the nearer one. Like a magnifying glass, the nearer galaxy bends the light from the more distant one, creating a white spot or 'Einstein ring', named for the physicist who predicted this gravitational effect.

ISAAC NEWTON 1643–1727
A controversial genius

The son of an illiterate yeoman farmer and an aristocratic mother, Isaac Newton hardly seemed predestined to become a great scholar. Yet he would revolutionise the fields of physics, optics and maths, becoming one of science's most acclaimed geniuses. A solitary man, he was gnawed by ambition, sensitive to criticism and at times was tormented, possibly even paranoid. He remains an enigmatic figure.

'Sir, Some time after Mr. Millington had delivered your message, he pressed me to see you the next time I went to London. I was averse; but upon his pressing consented, before I considered what I did, for I am extremely troubled at the embroilment I am in, and have neither ate nor slept well this twelve month, nor have my former consistency of mind … [I] am now sensible that I must withdraw from your acquaintance, and see neither you nor the rest of my friends any more, if I may but leave them quietly …'

Dated September 13, 1693, this letter was addressed to the diarist Samuel Pepys by his one-time companion Isaac Newton. The evidence it presents of a troubled mind is all the more striking as it seems that Pepys never asked Millington to contact Newton, nor sought in any other way to meet up with him.

Psychosis and delirium

At the time, the 50-year-old Newton was going through a particularly dark period in his life. Recent research has suggested that he may have suffered for several months from a persecution complex accompanied by hallucinations. One author has claimed that Newton's problems may have been caused by mercury poisoning brought on by the alchemical experiments he had long been in the habit of conducting.

It is hard to judge the truth of such speculation today, but it does appear that there were several Isaac Newtons. There was the solitary truth-seeker, antisocial, hypersensitive and unstable. There was the political operator, an expert and at times unscrupulous strategist.

Lasting work
A lead bust of Sir Isaac Newton, president of the Royal Society from 1703 to 1727, created by the sculptor John Cheere (1709–87).

NEWTON THE ALCHEMIST

Newton practised alchemy for 30 years, and one in ten of the books in his library was on the subject. Although this obsession might seem strange today, it was natural enough at a time when alchemy had yet to be discredited. Alchemical procedures, such as filtering and distillation, prefigured the techniques of modern chemistry, which did not then exist as a separate science.

Young Newton's sundial
The engraving below is said to be a copy of a sundial that Newton drew on the walls of Woolsthorpe Manor, his boyhood home.

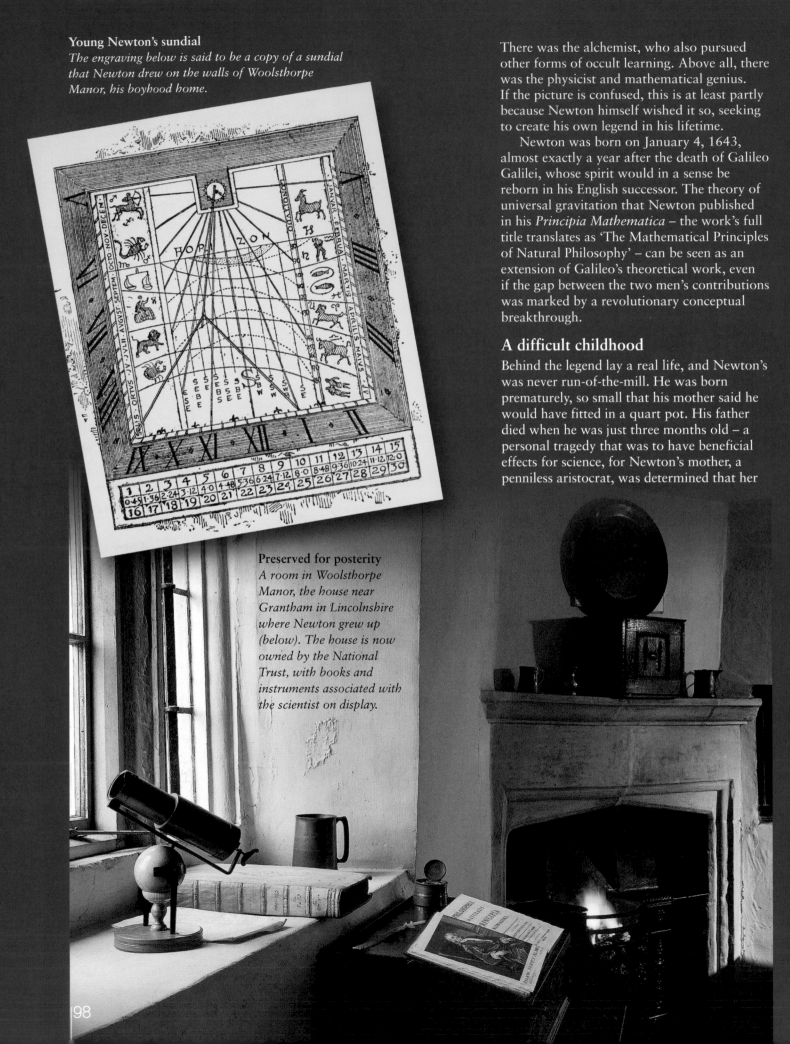

Preserved for posterity
A room in Woolsthorpe Manor, the house near Grantham in Lincolnshire where Newton grew up (below). The house is now owned by the National Trust, with books and instruments associated with the scientist on display.

There was the alchemist, who also pursued other forms of occult learning. Above all, there was the physicist and mathematical genius. If the picture is confused, this is at least partly because Newton himself wished it so, seeking to create his own legend in his lifetime.

Newton was born on January 4, 1643, almost exactly a year after the death of Galileo Galilei, whose spirit would in a sense be reborn in his English successor. The theory of universal gravitation that Newton published in his *Principia Mathematica* – the work's full title translates as 'The Mathematical Principles of Natural Philosophy' – can be seen as an extension of Galileo's theoretical work, even if the gap between the two men's contributions was marked by a revolutionary conceptual breakthrough.

A difficult childhood

Behind the legend lay a real life, and Newton's was never run-of-the-mill. He was born prematurely, so small that his mother said he would have fitted in a quart pot. His father died when he was just three months old – a personal tragedy that was to have beneficial effects for science, for Newton's mother, a penniless aristocrat, was determined that her son should be educated. He therefore became the first in the Newton line of yeoman farmers to learn how to read.

When Newton was only three, his mother remarried and went to live with her new husband, a clergyman Reverend Barnabus Smith, leaving young Isaac in the care of his grandmother in the family home, Woolsthorpe Manor. She came back seven years later following Smith's death, but the intervening period of abandonment had a lifelong effect on Newton. His mother's death in 1693 may have been a factor in plunging him into depression.

Newton attended classes at King's School in Grantham, where his signature can still be seen carved into a library windowsill. A sickly, neurotic youth, he already showed a passion for making small objects, perhaps honing the talents that would be put to good use in the 1670s, when he constructed the first reflecting telescope and created prisms that split beams of light into their constituent colours. It was during his later schooldays that he had his only known romantic attachment with a member of the female sex, a certain Anna Storer. Later, his apparent misogyny and close personal relationship with a young disciple, the Italian mathematician Fatio de Duillier, sparked speculation that he was homosexual.

The year 1660 marked a turning-point for Newton. His mother had taken him out of school the previous year in the hope of persuading him to run the family farm like his father before him. The 17-year-old was deeply unhappy at the prospect and his schoolmaster

A THEORY OF COLOUR

According to the popular account, in 1672 Newton used a prism to demonstrate that so-called 'white' light is formed of all the colours of the rainbow, thereby establishing the modern theory of colour. In fact, few people at the time were persuaded by his theory. Robert Hooke took issue with Newton, insisting that the prism somehow tinted the light passing through it. The result was a fierce controversy, but one that Newton finally won. His work on light marked the culmination of a tradition of optical research by Arab and Islamic scholars. As early as 1021 the Persian Ibn al-Hasan, better known in the West as Alhacen, established the fundamental principle that a ray of light is composed of a flow of particles whose behaviour obeys geometrical rules. In 1637 René Descartes rejected the notion of a flow of particles, preferring to view light as a wave. Newton opted for the particle theory and his corpuscular model prevailed for some time. The debate prefigured that of Planck and Einstein in the 20th century, which eventually established that light exhibits properties both of particles and of waves.

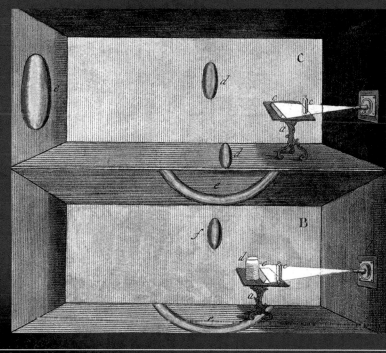

Separating the spectrum
An illustration from a German science manual published in 1800, illustrating Newton's optical experiments (above).

Another view of Newton
The visionary English artist William Blake was a critic of the Scientific Revolution of which Newton was such an important part. Blake's representation of the scientist (left) shows him engaged in an heroic but ultimately doomed attempt to impose order on nature.

persuaded his mother to let him return to school. From there, he went on to Trinity College, Cambridge, where he studied not just Aristotle but also the works of Copernicus, Galileo, Kepler and Descartes, acquiring the grounding for his later career.

Two exceptional years

The most productive years of Newton's life were 1665 and 1666, when he returned to Woolsthorpe to escape the ravages of plague then sweeping London and Cambridge. It was during this period he formulated the basic concepts that would inform his future theories. In physics, he concentrated on optics and worked on the idea of universal gravitation. In maths he laid the foundations of what he called the 'method of fluxions', now known as infinitesimal calculus (see panel below), which would also prove a vital tool for other disciplines, physics among them.

Newton came to dominate the scientific stage. His work on light and optics, published between 1672 and 1704, as well as on gravitation and on calculus all confirmed his central role. In 1703 he was made life president of the Royal Society and its journal, *Philosophical Transactions*, became one of his principal mouthpieces. Two years later he was knighted. His middle years were blighted by controversy, but the polemics cooled as his rivals died or moved on. From 1700 to 1727 he also served as Master of the Royal Mint.

In March 1727 Sir Isaac presided over his last session of the society. By the end of the month he was dead. He was laid to rest in Westminster Abbey, where his imposing monument bears a eulogistic inscription. Earlier, he had provided his own, surprisingly modest epitaph, writing: 'I do not know what I may appear to the world, but to myself I seem to have been only like a boy playing on the sea-shore, and diverting myself in now and then finding a smoother pebble or a prettier shell than ordinary, whilst the great ocean of truth lay all undiscovered before me.'

Treasury of knowledge
Located in Trinity College, Cambridge, Newton's alma mater, the Wren Library (above) preserves Newton's own annotated copy of the Principia Mathematica.

Newton's contemporary
Among many other achievements, Gottfried Wilhelm Liebniz (1646–1716, right), the German philosopher and mathematician, was the first man to spell out in detail the potential of the binary system as used in modern computers.

LEIBNIZ – AN UNWITTING RIVAL

From 1666 on, Newton worked on a new method of calculation, known today as infinitesimal calculus. For whatever reason, he chose not to publish anything on the subject until 1693. Meanwhile, the German philosopher Gottfried Leibniz had devised a similar system, using a different notation, and had published details as early as 1684.

From 1699, members of the Royal Society (of which Newton was then president) launched a sustained attack on Leibniz, accusing him of plagiarising Newton's ideas. Modern historians believe each came up with the discovery independently, but at the time the 'calculus controversy' cast a shadow over both men's lives.

Champagne c1695

The champagne method was born of a combination of luck and physics. Because of the comparative harshness of the climate, grapes in the Champagne region of France in the 17th century were picked before they reached maturity. They were used to make acidic wines that were low in alcohol and had to be bottled early to conserve their flavour. The cold halted the fermentation process, which then restarted with the coming of spring, causing bubbles to form. The sparkling wines thus accidentally created became popular, particularly at the court of France's Sun King, Louis XIV, but the degree of effervescence varied from one bottle to the next and also changed with the seasons. Some bottles even exploded under the pressure of the gas they contained.

Controlling the process

The first attempts to control the fermentation process in Champagne wines date from the end of the 17th century when new production methods were introduced, notably the addition of sugar and yeast to encourage a second fermentation. Another innovation was riddling (agitating) and uncorking the bottles to force out impurities. From that time on, the process has been steadily refined until today the production of a good vintage requires no fewer than 11 separate operations.

An aristocrat among drinks
An engraving in the style of French artist Antoine Watteau shows a fashionable lady drinking champagne in 1780. In the early years the wine was sweeter than it is today; the designation brut, *indicating a very dry wine, was introduced for the British market in 1876.*

DOM PÉRIGNON – INVENTOR OF CHAMPAGNE?

Pierre Pérignon (1638–1715) has traditionally been considered the father of champagne. He was a monk in the Benedictine Abbey of Hautvilliers, north of the town of Épernay in the Champagne region, where he served as the cellarer with responsibility for the monastery's supplies of food and drink, including the cultivation of grapes. The story goes that one year he stoppered the bottles with beeswax instead of the wooden stoppers that had previously been used. After a few weeks the bottles exploded, for the sugar in the wax had acted on the wine to create a second fermentation, making the liquid effervescent. In response, Dom Pérignon began using thicker glass bottles fastened with corks. But the truth is that by then similar methods had already been used in England to bottle wine imported in barrels – from the Champagne region.

A BRITISH DISCOVERY?

Historians now believe that the first person to advocate the use of sugar to make sparkling wine was a British doctor, Christopher Merret, in a paper presented to the Royal Society in 1662.

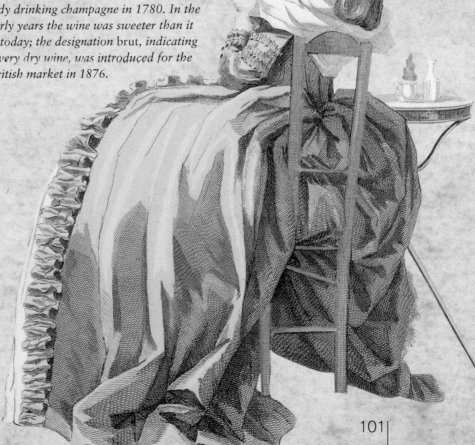

ADVANCES IN ANATOMY AND MEDICINE
A new view of the human body

For two millennia ideas about the human body were largely based on the key texts of Classical antiquity, but from the early 17th century those perceptions began to change. Experiment and observation became the order of the day, pushing the boundaries of knowledge into previously unexplored territory.

Four humours, four temperaments
An illustration from a 15th-century manuscript depicts the personality types associated with the four humours of classical medical theory: melancholic (associated with black bile), phlegmatic (phlegm), sanguine (blood) and choleric (yellow bile). Each one provides elements of the healthy individual (centre), believed to exhibit a balanced mixture of all four.

Until the start of the Renaissance, the vision of the physical world that prevailed among Europeans, whether scholars or laymen, was drawn from a basket of ideas inherited from antiquity and approved by the Church. The first of these accepted opinions to be seriously challenged was the notion that the Sun moved around the Earth. Preconceptions about the functioning of the human body were next in line. Long-established views crumbled as much of what people had traditionally taken as truth was shown to be false.

The body, it now turned out, was not composed of the four humours – blood, phlegm, black bile and yellow bile – described by Hippocrates in the 5th century BC. Galen's reworking of the canon of Classical medicine six centuries later was revealed to be no more reliable. Direct observation was debunking these most cherished traditional theories. The result was a growing appetite for accurate information, even if those seeking the facts risked being accused of heresy or madness.

The birth of a science

A necessary preliminary step to rebuilding medicine on the basis of direct, observable evidence, was to challenge the ban on anatomical dissection. Andreas Vesalius is the best-known of the early anatomists, sometimes known as the father of modern anatomy. In his *De humani corporis fabrica* ('On the Structure of the Human Body'), published in 1543 in Calvinist Switzerland, he presented ground-breaking new descriptions of the human skeleton, viscera, muscles and nervous system, revealed in 300 detailed woodcuts. The book was an immediate success and helped to disseminate Vesalius's findings among Europe's educated classes.

In the late 16th century most of the new knowledge about human anatomy came from Italian universities, notably Padua where Vesalius had been a professor at the School of Anatomy. His successors there duly followed in his footsteps. Renaldus Columbus, Gabriel Fallopius and Hieronymus Fabricius each pursued independent lines of enquiry. Columbus studied the heart and circulation, paving the way for the future work of William Harvey. Fallopius studied the female genital organs, describing the ducts leading from the ovaries to the uterus that are known today as

PIONEERS OF HUMAN ANATOMY

Up until the 16th century it was necessary to obtain permission from Church and State authorities before dissecting a human corpse. For want of such permission, scientists often used animals, which were not considered to have souls. The result was the survival of a number of unfounded superstitions – for example, that men had one rib fewer than women who, according to the Bible, had been created from Adam's rib. In the 17th century, the study of the body's structures gradually developed into a science and anatomy lessons became almost fashionable. Indeed, the ending of the old taboos encouraged excesses and crowds flocked to the amphitheatres to gawp at naked cadavers, which were often those of freshly executed criminals.

The Italian anatomist Giovanni Battista Morgagni was one of the first people to compare the anatomy of dissected corpses with the clinical symptoms exhibited by invalids and knowledge gained from post-mortems. In 1761 he published *De sedibus et causis morborum per anatomen indagatis* ('Of the Locations and Causes of Diseases Investigated through Anatomy'),

thereby founding the new medical discipline of anatomical pathology.

One result of the popularity of anatomical studies was the production of illustrative wax mannequins. From the 19th century, formaldehyde was increasingly used to help preserve corpses. By then, public lessons had been banned and dissection banished to the laboratories.

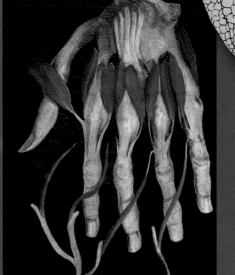

Delving under the skin
A public dissection of a human corpse, performed in a Renaissance operating theatre (top). The Italian anatomist Marcello Malpighi concentrated on lungs; his sketch above shows the lungs of a frog. The depiction of a human hand (left) is from an anatomical table prepared by Hieronymus Fabricius, who founded the anatomical theatre at the University of Padua in 1584.

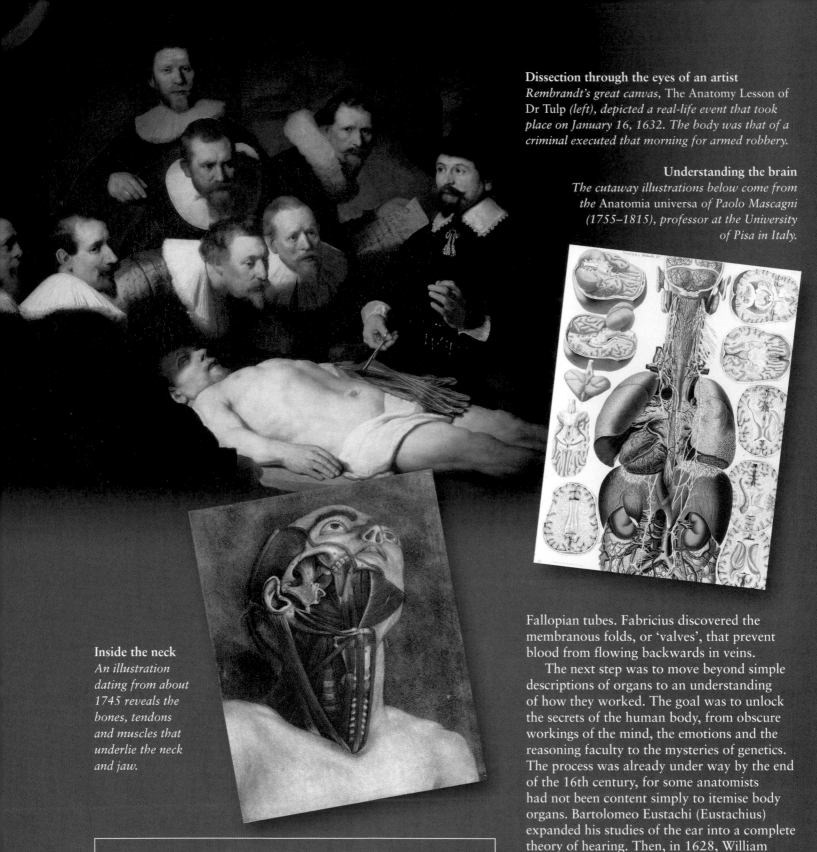

Dissection through the eyes of an artist
Rembrandt's great canvas, The Anatomy Lesson of
Dr Tulp *(left), depicted a real-life event that took
place on January 16, 1632. The body was that of a
criminal executed that morning for armed robbery.*

Understanding the brain
*The cutaway illustrations below come from
the* Anatomia universa *of Paolo Mascagni
(1755–1815), professor at the University
of Pisa in Italy.*

Inside the neck
*An illustration
dating from about
1745 reveals the
bones, tendons
and muscles that
underlie the neck
and jaw.*

A NEW CURE FOR FEVERS

Europe acquired a new tool to combat fevers when quinine bark was
brough back from South America, where it had long been used by
the indigenous peoples of Peru. In Europe the drug became known as
'Jesuit powder' because of the role the missionaries played in its
transmission. It was ineffective against maladies like smallpox, but it
quickly proved its worth against recurrent sicknesses of the type known
since Hippocrates' day as quotidian, tertian or quartan fevers, depending
on the number of days between attacks. Such intermittent bouts are
a symptom of malaria, which was common in Europe at the time.

Fallopian tubes. Fabricius discovered the
membranous folds, or 'valves', that prevent
blood from flowing backwards in veins.

The next step was to move beyond simple
descriptions of organs to an understanding
of how they worked. The goal was to unlock
the secrets of the human body, from obscure
workings of the mind, the emotions and the
reasoning faculty to the mysteries of genetics.
The process was already under way by the end
of the 16th century, for some anatomists
had not been content simply to itemise body
organs. Bartolomeo Eustachi (Eustachius)
expanded his studies of the ear into a complete
theory of hearing. Then, in 1628, William
Harvey resolved the great physiological
enigma of the day, establishing the essential
principles of the circulation of the blood. His
explanation of how the heart and its vessels
worked marked the point at which physiology
became a science.

Anatomy through the microscope

The invention of the microscope early in the
17th century was a major step forward. The
instrument exposed previously invisible detail,

Alchemist at work
An engraving from Annibal Barlet's Vray et méthodique cours de la physiqe résolutive vulgairement dite chymie *(1653) shows the author demonstrating alchemical techniques to students.*

enabling scientists to investigate hidden nooks and crannies. Using a microscope Anton van Leeuwenhoek discovered red corpuscles, while Marcello Malpighi produced close-up studies of plant and animal tissues, thereby claiming for himself the title father of histology.

One result of such work with microscopes was a re-opening of the debate about the origins of life. Confirming Harvey's intuitive belief that human beings, like animals, developed from ova or eggs, in 1672 a Dutch scientist named Regnier de Graaf produced the first microscopic evidence of ovarian follicles. Soon afterwards van Leeuwenhoek described spermatozoa. The debate thereafter centred on the way in which spermatozoa and ova interact to create the embryo, a question that would not be resolved for two centuries. Many favoured the theory of 'spontaneous generation', which proposed that male sperm contained tiny homunculi that the female egg and uterus served to feed and house.

The body as a laboratory

Two conflicting views of the human body – one from the iatrochemists, the other from iatrophysicists – struggled for supremacy. The former believed in the central importance of chemical processes. A leader of this school of thought, with significant influence on the development of medicine in Europe, was Jan Baptist van Helmont of Brussels. Van Helmont was particularly concerned with digestion, which he believed was a chemical process involving six types of fermentation; in his view the digestive juices acted like acid on a leather glove. His younger Dutch contemporary Franciscus Sylvius asserted that fermentation, effervescence and putrefaction were the three universal chemical processes in physiology. Another fermentation proposal came from an English doctor, Thomas Willis, who viewed the human body as a distillery in which chemical reactions were responsible for digestion, muscular contraction and the transmission of messages by the nervous system.

By the end of the century such views were being challenged by the mechanistic theories of the iatrophysicists. Descartes was a precursor of this approach, stating in his posthumously published *Traité de l'homme* ('Treatise on Man'): 'I take the body to be nothing more than a statue or an earthly machine.' Galileo

A CRITICAL VOICE

Although the 17th century saw the growth of scientific medicine, the profession was not transformed overnight. Despite the progress made, most day-to-day treatment was still based on the 'four humours' model inherited from the ancient world, remaining reliant on purgatives and blood-letting.

For the great French playwright Molière, the doctors of his day were for the most part charlatans – respectable on the surface but in reality more interested in making money than in curing their patients. In his 1673 play *Le malade imaginaire* ('The Imaginary Invalid'), a doctor suitably named Purgon has nothing to offer his patient but enemas.

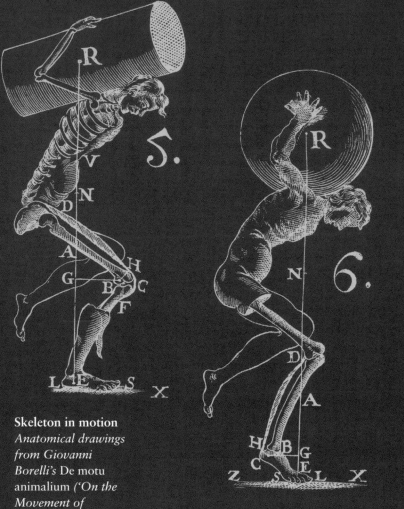

had used mechanical laws to describe the articulation of limbs, to muscle movements and to walking and balance, comparing such motions to the action of levers, pulleys and weights suspended on ropes. He believed that physiological processes – ranging from muscular contraction and secretion of glands to the operation of the heart, brain and lungs – were all controlled by physical laws. Galileo's view had a significant effect on later research. His pupil Giovanni Borelli, for example, analysed the motion of human and animal limbs using the laws of statics, a branch of mechanics.

Animism and vitalism

The quarrel between the chemists and physicists ran out of steam in the course of the 18th century, as medical researchers came to recognise that there was validity in both points of view. In his studies of the digestive system, for example, Hermann Boerhaave, the Dutch pioneer of teaching hospitals and clinical medicine, discovered both mechanistic elements in the breakdown of foodstuffs in the stomach and chemical ones in the process of fermentation. For Boerhaave the activities of the human body, which is composed of solids and liquids, represented life; death ensued when these activities came to a halt. Illness

Skeleton in motion
Anatomical drawings from Giovanni Borelli's De motu animalium *('On the Movement of Animals') of 1680.*

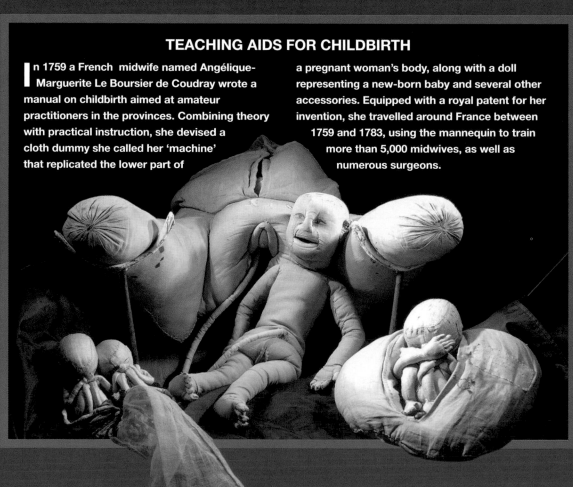

TEACHING AIDS FOR CHILDBIRTH

In 1759 a French midwife named Angélique-Marguerite Le Boursier de Coudray wrote a manual on childbirth aimed at amateur practitioners in the provinces. Combining theory with practical instruction, she devised a cloth dummy she called her 'machine' that replicated the lower part of a pregnant woman's body, along with a doll representing a new-born baby and several other accessories. Equipped with a royal patent for her invention, she travelled around France between 1759 and 1783, using the mannequin to train more than 5,000 midwives, as well as numerous surgeons.

THE PLAGUE – A CONTINUING SCOURGE

Despite the advances in medicine, doctors remained powerless against infectious diseases. The plague had decimated Europe in the 14th century and outbreaks continued to occur, notably in London in 1665 and in Marseille in 1720, when an epidemic killed a third of the city's population, some 50,000 people. At the time medical opinion held that plague was carried by noxious vapours. The bacterium responsible was not identified until 1894; the method of transmission, via rat-borne fleas, was demonstrated four years later. DNA analysis of medieval corpses has now confirmed that this was indeed the pathogen that caused the plague outbreaks of the past.

Plague in Marseille
Michel Serre painted this image of the epidemic that afflicted the city, killing more than 1,000 people a day at its peak.

resulted from imbalance between the two – a theory that recalled certain aspects of the Hippocratic system. The German physician Friedrich Hoffmann also held that 'movement is everything', but in addition he insisted on the importance of a substance in the brain he called 'nervous ether', which he believed to be a subtle fluid formed in the nerves that made the heart beat and the muscles contract.

Hoffmann's compatriot George Stahl introduced the idea of an immaterial unifying principle, the *anima* – a sentient soul that regulated the body's mechanisms and kept death at bay. Such ideas spread, particularly in northern Europe. Stahlism's French equivalent was vitalism, propagated by Théophile de Bordeu and Paul Joseph Barthez. This held that the body was animated by a life-force – dubbed the 'vital principle' by Barthez – whose deterioration was said to cause sickness. Such views were also held by the London experimentalist and pioneer surgeon John Hunter. These different theories of the human body and its functions gave birth in turn to fresh ways of classifying illnesses and determining the choice of treatment.

Birth of a discipline

Enriched by the flow of new opinions, modern physiology finally got off the ground in the course of the 18th century. Experimentation was by then the order of the day. Reforms affecting the teaching and practice of medicine followed, spreading the scientific approach through the profession. Further progress in microbiology (notably by Louis Pasteur), in experimental medicine and in the study of the nervous system all helped to bring about a new relationship between the doctor and the patient. Physicians took an increasingly hands-on attitude towards healing their patients. The era of modern medicine had begun.

Open heart
At the Children's Academy in the German city of Fulda, visitors can explore inside a huge blown-up model of a human heart. By following the route taken by red corpuscles, they learn in the process about the circulation of the blood.

THE STEAM ENGINE – *c*1698

A revolutionary new source of energy

The invention of the steam engine in the late 17th century changed the course of history, leading to the industrial era. The first such machine was devised by a Frenchman named Denis Papin, who built on discoveries about vacuums and atmospheric pressure made in the preceding decades.

Demonstrating steam power
Denis Papin explains how his steam machine works to colleagues at the University of Marburg in Germany, where he taught from 1688 to 1695.

One day in 1712 a plain coffin was carried through the streets of London for burial in a common grave. Inside were the mortal remains of a largely forgotten scientific genius who had been pursued by creditors even to his death-bed. It was a sad end for the man who had discovered the working principle of the steam engine, one of the key developments of the early Industrial Revolution, which would trigger Britain's economic take-off from the late 18th century on.

Denis Papin had first developed an interest in steam in Paris in 1673, while working alongside the pioneering Dutch scientist Christiaan Huygens. Following in the footsteps of the Italian Evangelista Torricelli (who had invented the barometer in 1643) and the German Otto von Guericke (vacuum pump, 1650), the two men were experimenting with vacuums and atmospheric pressure.

Their researches involved driving a piston up and down a vertical cylinder. The shaft of the piston was attached to a rope connected via pulleys to a counterweight outside the cylinder. By setting fire to a small charge of gunpowder inside the casing they caused gases in the cylinder to expand, raising the piston and lowering the counterweight. When the pressure within the confined space was subsequently reduced the process went into reverse, so the piston fell and the weights rose. Papin described their work in a 1675 text with a title that translates as *New Experiments on Vacuums, with a Description of the Devices Used to Make Them.*

Unwanted inventions

Papin left Paris in 1675 and spent the next few years travelling around Europe. As a result it was only in the late 1680s that he again took up the work where he had left off, this time at the University of Marburg in Germany. In a fresh round of experiments he replaced the gunpowder with water that he heated to boiling point, demonstrating that, like the expansion and contraction of gas, the condensation of steam could drive the piston and so raise and lower the weights via pulleys. In so doing, he established the principle of the steam engine. He set out his findings in a 1687 pamphlet entitled *Description and Uses of a New Device for Lifting Water*. Three years later, seeking to put his invention to use, Papin published another essay, *A New Method of Exerting Great Force at Little Cost*, outlining various potential applications.

Papin was keen to put some ideas into practice himself. One of these was a boat that

Ill-fated steamboat
A 19th-century illustration of the destruction of Papin's steamboat by disgruntled boatmen on the River Weser in Germany in 1707.

Papin's pump
Made of copper and metal alloys, Papin's steam-powered pump for raising water weighed 38kg.

he had built powered by paddlewheels driven by a steam pump. But in September 1707, as he steered his steam-powered boat down the River Weser toward the port of Bremen, it was attacked and destroyed by boatmen fearful of the threat it posed to their livelihood.

He next turned his attention to a problem that had long plagued miners – the never-ending need to pump water out of mine shafts. Papin's solution took the form of a machine incorporating a boiler, a closed vessel and a reservoir of water. The idea behind it was that steam from water heated in the boiler would fill the reservoir. By then cutting off the steam and lowering the temperature of the reservoir with cold water from the second vessel, the steam would condense, creating a vacuum that would in turn suck up the water from the mine. In practice Papin's fire pump never got beyond the experimental stage for want of funds to develop the idea.

English inventors take up the baton

A Devon-born military engineer named Thomas Savery, who was acquainted with Papin's work, patented a steam pump designed along similar lines, describing it in a 1702 book entitled *The Miner's Friend*. Despite the fact that his machine had serious drawbacks,

DENIS PAPIN – UNLUCKY GENIUS

Denis Papin was born in August 1647 near Blois in northern France, the fourth of 13 children of a Protestant doctor. After studying medicine at the University of Angers, he travelled to Paris, where he worked with the Dutch scientist Christiaan Huygens, before moving to London to assist Robert Boyle. The revocation of the Edict of Nantes in Catholic France in 1685 ended an era of tolerance towards French Protestants, preventing him from returning to his home country. Instead, encouraged by the backing of Prince Charles Augustus of Hesse, he took up the chair of mathematics at the University of Marburg in Germany. Papin developed many inventions during his stay in Germany, but the destruction of his steamboat (see above) persuaded him to return to England. In the interim, however, all his old friends there had died. Papin passed away in 1712, in penury and forgotten. In his last known letter he wrote: 'I am in a sad way, for in trying to do good I only seem to make enemies. Yet I fear nothing, putting my trust in an all-powerful God.'

Savery obtained the exclusive right to produce these 'fire engines', as they were known, for 35 years. Unlike the unfortunate Papin, he profited handsomely from the devices, which marked the start of steam's industrial career.

In 1712 a Devon ironmonger named Thomas Newcomen, working with his partner the glassmaker and plumber John Cawley, came up with an improved design for the fire pump. Their device delivered 12 piston strokes a minute, raising 500 litres of water from a depth of 45 metres. They entered a partnership with Savery to produce the new pump under his patent and it was immediately successful not just in England, where it was eventually adopted by most of the 'wet' mines, but also across Europe. Yet there were still problems with the mechanism, which was heavy and expensive to run, consuming a phenomenal amount of fuel and often breaking down.

Steam pioneers
Thomas Savery's steam-pump (left) and Newcomen and Cawley's improved version (right). In about 1790, the Newcomen engine was adopted by the ironworks at Coalbrookdale in Shropshire (far right, top). With its coke-fired blast furnace, Coalbrookdale was an early producer of steam-engine cylinders and the crucible for the Industrial Revolution.

Perfecting the mechanism

Half a century passed before the Scots inventor James Watt finally found a way of decisively improving the design. While working at the University of Glasgow, Watt was called on to repair a Newcomen engine. He quickly realised that much of the heat it produced was lost in simply reheating the cylinder. To remedy the defect, he installed another chamber, separate from the piston cylinder, where the steam could condense, thereby reducing heat loss by some 75 per cent. In 1769 Watt took out a patent on his version of the machine, which not only used less fuel but also provided vastly

increased output. Thereafter he continued to make further improvements, introducing a double-acting engine that employed the steam itself to activate the piston and adding a control to adjust the piston's speed.

The spread of steam

With efficient machinery at last available, there was nothing to stop the spread of steam power in an England that was already setting itself up as the standard-bearer of free-market capitalism, committed to free trade and minimum state intervention. In 1786 Edmund Cartwright, a Church of England clergyman, harnessed steam to drive a power loom, and before long mechanical looms had spread throughout the textile trade. What had been a cottage industry was transformed into a well-capitalised enterprise fed by cotton and other raw materials from the colonies

JAMES WATT, HANDYMAN ENTREPRENEUR

Born on January 19, 1736, at Greenock near Glasgow, James Watt was the son of a shipwright. He was sickly as a boy and mostly schooled at home, but following family financial problems he was forced to move to London in 1854 to study instrument-making. Watt returned to Glasgow after a year and took up a job in charge of the maintenance of laboratory equipment at the city's university. There he became interested in steam engines, and soon found ways to improve their performance, notably by adding a separate condensing chamber to reduce heat-loss in the main cylinder. To commercialise his designs he teamed up with the entrepreneurs John Roebuck, who went bankrupt, and then Matthew Boulton, proprietor of the Soho Manufactory near Birmingham, with whom Watt enjoyed an immensely successful partnership. Between 1776 and 1800, the year of his retirement, Watt and Boulton produced some 500 steam engines. Appointed a Fellow of the Royal Society, Watt lived on until 1819, dying rich and famous at the age of 83.

Watt's steam engine
This working model was made in 1821, two years after Watt's death.

111

Corn threshers *A painting by Peter de Wint shows a steam-powered threshing machine at work in the early 19th century.*

Russian textile workers

A 1904 painting by N A Kassatkine shows workers at Orekhovo-Zuyevo, an industrial town east of Moscow famous for its textile industry. There were 17 factories there in 1890.

and employing a vast labour force sweating in factories. Steam-powered threshers followed, the first coming into operation in 1802. By the 1830s there were 15,000 of these machines in Britain, as compared with 3,000 in France and just 1,000 in Prussia.

In the first half of the 19th century steam was also put to use to speed up transport. The results were nothing short of revolutionary. Even before the development of passenger services, steam-powered locomotives were being used to shift huge volumes of coal, steel products and textiles on rails laid within industrial and manufacturing works. From Britain, the railway mania spread first to the

STEAM POWER TAKES TO THE ROAD

The mechanisation of land transport received a significant boost from Nicolas-Joseph Cugnot, a military engineer from Lorraine in eastern France. In the course of his work he came up with the idea for a three-wheeled cart designed to carry cannons which was equipped with a front-mounted boiler and a steering mechanism. A model weighing 2.5 tonnes, built in 1770, proved capable of moving at 4 km/h, but only for about 15 minutes at a time. It also proved unstable and the project was abandoned.

ANCESTOR OF THE PRESSURE COOKER

The prototype of the pressure cooker was a device known as the 'steam digester' invented by Denis Papin in 1679. A cast-iron cylinder, equipped with a safety valve, it worked on the principle that increasing pressure in a sealed container filled with water raises the temperature above water's normal boiling point to 115°C. This means that foodstuffs including the toughest meats cook quicker than in a normal pan, where the temperature never exceeds 100°C. Papin described his invention in *A New Digester or Engine for Softning Bones*, published in England, where he was living at the time.

The principle purpose of the device was to extract gelatin from animal bones and Papin demonstrated it to the Royal Society at the invitation of Robert Boyle, modestly describing its operation as 'a rather brutal method of cooking'. He also spent time exploring the preservative effects of the gelatin it produced, noting that it helped to delay spoilage in apples and gooseberries. Yet once again Papin failed to exploit his idea commercially, and another two centuries would pass before pressure cookers came into general use, equipped by that time with gauges to indicate the pressure building up inside . Modern pressure cookers are generally set to 15 psi above the surrounding atmospheric pressure, reducing cooking times by an average of 70 per cent.

The steam digester *Papin's device was a simple closed cylinder with a release valve.*

FROM STEAM POWER TO THERMODYNAMICS

By the turn of the 19th century steam engines were being produced in large numbers, especially in England, but no-one fully understood the physical principles involved in their operation. Scientists for the most part still held to the caloric theory, maintaining that heat took the form of a fluid that flowed from hotter to colder bodies, thus heating water and turning it to steam within the engines. The motive power was in their view an effect of this elastic caloric fluid, which built up the steam pressure on the piston. Lacking a theoretical basis for their work, the engineers who produced the steam engines had to establish the ideal maximum temperature for each machine on an ad hoc basis by trial and error. It was only after 1824 when French physicist Sadi Carnot produced the first accurate theory of a heat engine, thus laying the foundations of the second law of thermodynamics, that people came to understand that heat is a form of energy.

Continent and then around the world. Rail travel effectively redefined people's notions of distance, opening up fresh horizons, and spurred the growth of industrial capitalism by stimulating the production of coal, steel and cast iron, as well as of the rolling stock itself. Metal-hulled steam ships also played a vital part by bringing fresh capacity and more reliable schedules to maritime commerce.

In the decades after 1850, the primacy of steam would eventually be challenged by the arrival of electricity and the petrol-fuelled combustion engine, which between them provided the motive force for a second industrial revolution. Yet even though steam's own importance declined, the message remained 'Full steam ahead'.

Steam trains in China
China's rail network extends over 43,500 miles, making it the fourth largest in the world after the USA, Russia and Canada, with India coming fifth. On much of this network China still uses steam trains, although this photograph was taken in the 1920s.

113

An efficient tool for sowing grain

At the start of the 18th century English agriculture entered a new era of prosperity, thanks at least in part to the mechanisation of sowing. The introduction of Jethro Tull's seed drill signalled the coming of modern agricultural machinery and encouraged the transformation of farming from a subsistence activity into a large-scale commercial enterprise.

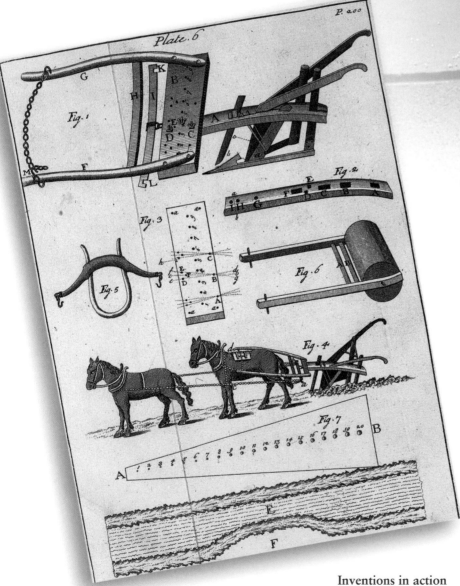

Inventions in action
A detailed illustration of new horse-drawn implements from an edition of Jethro Tull's The New Horse Hoeing Husbandry *published in 1743.*

Even though he is now considered one of the founding fathers of modern agronomy, Jethro Tull initially had no vocation for farming. Born in Berkshire, he attended St John's College in Oxford before being called to the Bar in 1699. He was heading for a political career at the time, but a lung complaint forced him to give up his place at Gray's Inn and travel south in search of a cure. Returning to England some months later, he settled down on the family estate in Berkshire and set about turning himself into an enlightened landowner, committed to advancing the rational agriculture that was then coming into vogue.

First steps

In England, as throughout Europe, people sowed seed by manually casting it onto soil that was rarely given any advance preparation. The broadcast seed fell as the breeze took it, sometimes too close together to grow into healthy plants, at others scattered too sparsely to make best use of the ground. Seeds were eaten by birds and damaged by heat and frost. Concerned by the wastage, Tull tried and failed to persuade his sowers to change their working habits. Frustrated, he decided instead to build a machine to do the job properly.

He unveiled his first horse-drawn seed drill in 1701. The device performed three functions. At the front a ploughshare dug triple furrows. A hopper in the centre dispensed the seed through a grooved rotating cylinder, depositing it at the bottom of the furrows. A harrow at the back closed the soil over the seed.

ENGLAND'S AGRICULTURAL REVOLUTION

In the course of the 18th century English agriculture moved from a subsistence activity into a fully-fledged commercial enterprise. There was an upsurge in the practice of enclosing land, which was suddenly commercially viable. Fallow fields and common pasture went by the board, as the new norm became hedged or fenced fields with named owners. The custom of rotating crops – planting different seed with different needs in successive years so as not to exhaust the soil – was imported from the Low Countries. Cereals such as wheat or barley alternated with crops for fodder, such as turnips or clover. These were mostly fed to cattle, which in turn provided milk and manure to enrich the soil. At the same time marshy land was drained and steps were taken to increase soil productivity. Machines were increasingly used to cultivate fields in regular furrows. The result was a doubling in wheat production between 1650 and 1800, although at the cost of growing social inequality that forced landless peasants to become agricultural labourers or migrate to the cities.

Green and pleasant land
A patchwork of enclosed fields in Cornwall, sweeping down to the sea (above). An engraving of 1858 (below) shows a steam engine equipped with a rotary tiller.

This first machine was not a success, whether because it was badly designed or because unwilling labourers deliberately misused it is not known. Whatever the case, Tull withdrew in 1709 to lick his wounds on Prosperous Farm, a family estate near Hungerford. Two years later he set off once more for southern Europe, where he spent three years observing agricultural methods, discovering among other things that Italian farmers were already using drills not very different from his own.

A profitable voyage

On his return in 1714, Tull improved his tools and refined his ideas. He developed a modified drill with a plough that dug deeper furrows, and spent much time and thought on developing ways of preparing the soil. He first published his ideas in 1731 in a book entitled *The New Horse Hoeing Husbandry, or An Essay on the Principles of Vegetation and Tillage*, arguing the case for pulverising the soil between furrows to release the nutrients within it. The work became a key text of the vast movement of agricultural improvement that was sweeping through English farming at the time.

Tull was not the first person to devise a seed drill, but his machine was nonetheless revolutionary as the earliest one to be driven by wheels. He got the idea from the inner workings of organs, which he had learned to play in his student years. It was the moving parts above all that made Tull's invention the first true agricultural machine, serving as a prototype for the seed drills that spread across Europe in the following century. By then the world of agriculture was ripe for mechanisation and set to become a major field of innovation and commercial activity.

EDMUND HALLEY 1656–1742
A universal talent

An astronomer, meteorologist, geographer and oceanographer as well as a competent engineer, Edmund Halley was passionately curious about many fields of science, but he owes his lasting fame to the comet that still bears his name. Halley worked out both the comet's trajectory and its periodicity, establishing that it appears in the Earth's skies once every 75 or 76 years.

Astronomer to the King
A portrait of Edmund Halley (top) shows him at the grand old age of 80. The map of stars, showing some of the southern constellations, is a detail from a globe that Halley prepared for George III in 1766.

Edmund Halley hailed from a prosperous middle-class background. He quickly caught the attention of his professors at Queen's College, Oxford, where he excelled at maths and the natural sciences. He developed an early passion for astronomy; when he was just 19 he wrote a letter to John Flamsteed, director of the Royal Observatory in Greenwich, pointing out numerical errors in the ephemerides (daily location tables) for Jupiter and Saturn. Flamsteed was so impressed that he invited the young man to join the Royal Society.

Eager for adventure, Halley set off two years later for the island of St Helena in the South Atlantic with the aim of drawing up the first map of the southern skies. There was genuine need for one at the time, for sailors still had no way of fixing longitude and were reliant on the stars to establish their position in the oceans. Halley took a particular interest in longitude and besides spending two decades working on his sky map and on various astronomical tables, he also devised new techniques of cartographic projection.

A star from nowhere

On December 18,1680, Halley had crossed the Channel and was on his way to Paris when he noticed an unfamiliar star with a curious long tail of light moving through the night sky. Arriving in the French capital six days later, he prolonged his stay to observe the comet from the Royal Observatory there, in company with its director, Jean-Dominique Cassini. He listened with fascination as Cassini expounded a new theory that comets orbited the Sun in the manner of planets, returning periodically to pass close by the Earth.

On his return to London, Halley immersed himself in ancient texts, drawing up a list of all the comets mentioned in the annals. Three of the reported appearances seemed to show

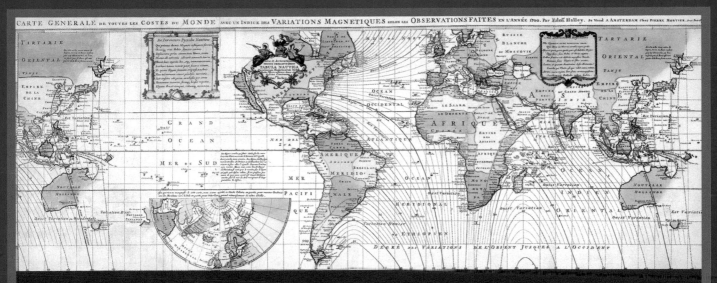

CARTE GENERALE DE TOUTES LES COSTES DU MONDE AVEC UN INDICE DES VARIATIONS MAGNETIQUES SELON LES OBSERVATIONS FAITES EN L'ANNÉE 1700. Par Edm. Halley. Se Vend A AMSTERDAM chez PIERRE MORTIER.

A PIONEER IN MANY FIELDS

In 1686 Halley published the first meteorological map of the world, showing the prevailing winds affecting the oceans. In 1690 he devised a diving bell for underwater research that was an early precursor of the bathysphere. In 1693 he drew up a pioneering actuarial table, based on demographic data prepared for the Silesian city of Breslau (now Wroclaw in Poland), predicting the percentage of new-born babies who would still be alive at given ages. His work was of service to the government, enabling civil servants to work out appropriate premiums for recently introduced life annuities.

Magnetic variations
Halley was the first person to measure and record magnetic declinations indicating the divergence between local magnetic fields and true north, as shown on this map of 1701.

distinctly similar characteristics: those of 1531, 1607 and a third in 1682, which he observed by telescope personally two years after his Paris stay. All three seemed to follow a course approximately the same distance from the Sun; their orbit was inclined at the same angle to the rest of the solar system and, intriguingly, the co-ordinates of their perihelia – the points where they most closely approached the Sun – seemed to be almost identical.

Newton, whom Halley met in 1684, lent his support to the younger man's work. Halley had been hesitating to link the three, despite all the evidence suggesting they might be one and the same, because there was a variation of a few months in the periodicity of their appearance in the heavens. Newton encouraged him to calculate the gravitational perturbations the comets might be

exerting themselves. Halley in fact went one better and was able to establish that the mass of Jupiter and Saturn was affecting the trajectory of the bodies he was studying. If allowance was made for their influence, the orbits of the comets of 1531, 1607 and 1682 were indeed identical.

Hale-Bopp – a brighter comet
First spotted on July 23, 1995, by two separate amateur astronomers – Alan Hale in New Mexico and Thomas Bopp in Arizona – the Hale-Bopp comet is 1,000 times brighter than Halley's comet seen from the same distance.

The orbits of comets
A diagram from the Mathematical Atlas *of German astronomer Tobias Mayer, published in 1745, showing comets orbiting the Sun (above).*

A TALENTED TRANSLATOR

In 1710 Edmund Halley produced a ground-breaking Latin edition of the *Conics*. This key work by the ancient Greek mathematician Apollonius of Perga gave the modern world, among other things, the words ellipse, parabola and hyperbola to describe conic sections of differing curves. The original work was divided into eight parts. The first four of these were available to Halley in Greek, but he had to translate Parts V, VI and VII from an Arabic manuscript, the only known version at the time. Part VIII was completely lost, so Halley compiled a synopsis from second-hand sources.

HALLEY AND NEWTON

One morning in 1684 Halley, then 28, visited Isaac Newton in his rooms at Cambridge. It proved a significant encounter. At the time, the mechanisms operating in the heavens were still largely unexplained. Copernicus had put the Sun at the centre of the universe, but no-one yet fully understood why or how the planets moved around it. Halley discussed the shape of the planets' orbits with Newton, and was astonished to hear the great man insist that they were elliptical. Three months later Newton sent Halley the calculations behind his view, and Halley thereby became the first person to learn of Newton's theory of the law of universal gravitation. Much excited by the revelation, Halley exhorted Newton to publish his findings and the result, three years later, was the *Principia Mathematica*, a key work in the history of science. Halley did everything in his power to promote the book, even paying the costs of the first edition.

Halley's Comet
On its last near approach to Earth, on March 12, 1986, the comet appeared brightest to observers in the southern hemisphere.

Halley's prediction

In 1705, in his *Synopsis of the Astronomy of Comets*, Halley made an extraordinary prediction: that the comet which had passed by the Earth in 1682 would return on Christmas Day 1758. Nobody paid much attention to his claim at the time, and it had been largely forgotten by the time of his death in 1742. But in 1757 a French mathematician, Alexis Clairaut, revisited Halley's calculations, which he then sought to verify with the help of astronomers from the Paris Observatory. Finding that Halley's figures held up, Clairaut encouraged astronomers around the world, however sceptical their attitude, to look out for the comet's reappearance. It was duly spotted in the skies above Dresden on the night of December 25, 1758.

The sighting was a posthumous triumph for Halley, and the comet was immediately named in his honour. Since he first studied it over London in 1682, it has returned four times. On its last visit, in March 1986, it was greeted by the space probe Giotto. The comet's next appearance is expected in February 2062.

The clarinet
c1700

A family of instruments
There are a dozen instruments in the clarinet family, ranging from the little soprano clarinet to the contrabass by way of the alto clarinet and basset horn.

T he clarinet began life as an adaptation rather than an invention. Its ancestry can be traced back to the medieval chalumeau, which had its heyday in the Renaissance. The chalumeau was a cylindrical tube made of cane or boxwood, provided with eight tone holes and equipped with a mouthpiece with a single reed vibrated by the player's breath. Its tonal range was limited: by pursing the lips hard, it was possible for a player to shift notes in the upper register by a full fifth, but as there was a gap between the two registers this technique was rarely employed.

At the end of the 17th century a German instrument-maker named Johann Christoph Denner determined to make some improvements. He modified the mouthpiece, bringing the reed into direct contact with the lips in order to improve the instrument's performance in the upper register. He also added two extra keys, including the twelfth, which enables players to move up an octave and a half and to pass instantaneously from one register to the other. One of Denner's sons, Johann David, is credited with replacing the chalumeau's straight end with a flared bell, which improved the production of low notes. The swelling bell reminded some observers of trumpets, in particular an early form of the instrument called the clarino, from which the clarinet duly took its name.

Later generations of instrument-makers made more changes that further improved the clarinet's range. In about 1815 a Russian named Ivan Müller devised a system of 13 keys controlling valves that blocked the tone holes the fingers could not reach. Then, in the mid 19th century, the German Theobald Boehm patented a system of mobile 'open rings', allowing a woodwind player to close several holes on an instrument at one and the same time with a single finger. The application of the Boehm system to clarinets produced the modern instrument, which has undergone only minor alterations since.

VIRTUOSO INSTRUMENT

T he clarinet has the largest variation in pitch of any woodwind instrument and lends itself particularly well to rapid playing. It has a remarkable range covering three octaves and a fifth as well as four registers, ranging from the lowest, known as the chalumeau, to the shrill altissimo. Clarinets are well suited to the works of the Romantic composers, as well as to jazz and to *klezmer*, a secular Jewish music popular at weddings and other celebrations.

Sidney Bechet (1897–1959)
The American jazzman was one of the greatest exponents of the clarinet. His composition Petite Fleur *became a million-seller in the 1950s.*

THE PIANOFORTE – c1709
The queen of musical instruments

Royal instrument
In May 1747 Johann Sebastian Bach performed before Frederick II of Prussia in the salon of the Sans Souci Palace at Potsdam (right).

The harpsichord was in its heyday when Bartolomeo Cristofori had the idea of modifying the inner workings so that the strings were struck rather than plucked. The instrument that he devised could be played softly (*piano*) or loud (*forte*), but initially it failed to win over either musicians or composers. Even so, craftsmen across Europe soon took up where Cristofori left off.

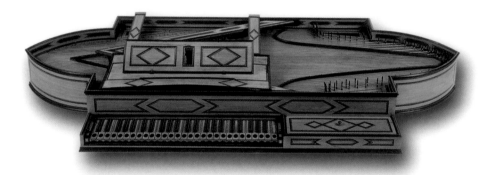

In his workshop in Florence, the instrument maker Bartolomeo Cristofori (1655–1731) put the finishing touches to a rosewood spinet constructed for the Grand Prince Ferdinand. Heir apparent to the Grand Duchy of Tuscany, the prince had brought Cristofori to the city from Padua in 1688. Yet talented and well regarded as he was, Cristofori was not satisfied with the results of his handiwork; to his ear, the spinets and harpsichords that he built lacked nuance in the playing. Striking the keys caused the strings to be plucked by sharp quills (plectra), that were encased in a mechanism called the jack. However hard or gently a player struck the keys, the sound remained essentially the same. Another stringed instrument, the clavichord, had overcome this defect, but the sound it produced was at best weak and metallic.

Taking the harpsichord as his model, Cristofori set out to create an instrument capable of more expression and tonal variation. He constructed a first prototype

AT THE COURT OF THE MEDICI

Bartolomeo Cristofori was employed as an instrument maker at the Medici court in Florence before becoming keeper of keyboard instruments to Grand Prince Ferdinand. He stayed on after the prince's death in 1713, becoming curator of all the royal instruments three years later.

Courtly music
Musicians employed at the Medici court by Prince Ferdinand, heir to the throne of Tuscany, who was himself a musician. The spinet (top) was made for the prince by Cristofori in 1693.

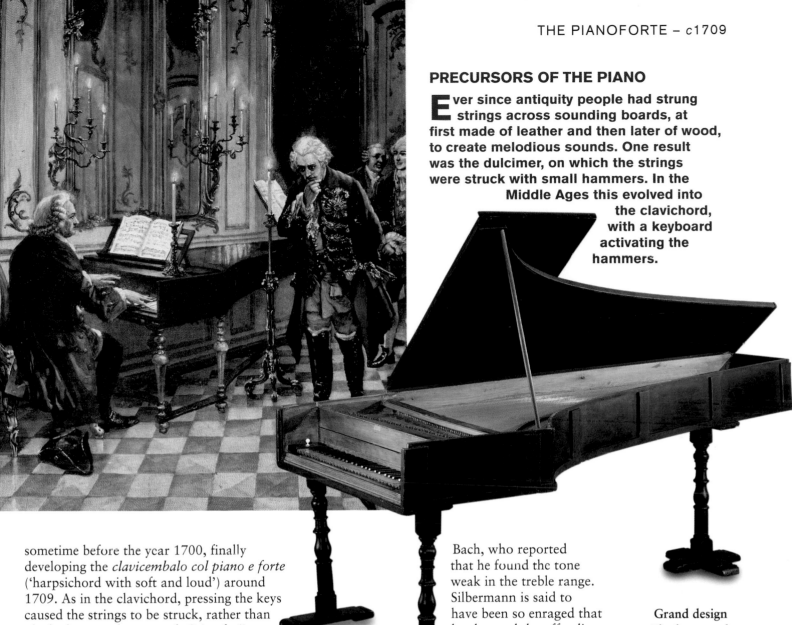

PRECURSORS OF THE PIANO

Ever since antiquity people had strung strings across sounding boards, at first made of leather and then later of wood, to create melodious sounds. One result was the dulcimer, on which the strings were struck with small hammers. In the Middle Ages this evolved into the clavichord, with a keyboard activating the hammers.

sometime before the year 1700, finally developing the *clavicembalo col piano e forte* ('harpsichord with soft and loud') around 1709. As in the clavichord, pressing the keys caused the strings to be struck, rather than plucked, to produce sound. Cristofori's principal innovation was to use articulated hammers that rebounded immediately after striking the strings, allowing the notes to continue vibrating. He continued to improve his invention over the following decades, making around 20 of the new models in the years up to 1726.

Ahead of his time

Yet for all his efforts, Cristofori's brainchild was not a success. A composer named Lodovico Giustini wrote pieces for the instrument from 1732 on, but in general composers had difficulty working out how to take advantage of its tonal dynamics, while performers struggled to adapt their techniques. The sound quality of the first pianos was thinner than that of pianos today, although the touch was if anything more precise.

An organ-maker named Gottfried Silbermann sought to remedy these defects, building the first German fortepiano, as the instruments were then called, in about 1726. He showed his handiwork to Johann Sebastian Bach, who reported that he found the tone weak in the treble range. Silbermann is said to have been so enraged that he chopped the offending piano up with an axe. Silbermann did not give up, however, and in 1747 he persuaded Frederick II of Prussia to buy several more of his instruments. Bach was invited to try these out also, and this time he approved of what he heard. Bach's belated endorsement served the piano well. Many were built in the second half of the 18th century, although little was standardised about the way they were made or the sound they produced.

Grand design
The first grand piano built by Cristofori (above) had 49 keys and a range of four octaves.

Square piano
Square pianos were popular from the 18th to the early 20th centuries. This model was built by Johannes Zumpe in 1767, with a mahogany frame and keys of ebony and ivory.

Virtuoso piano performers
The frontispiece from a Mozart concerto 'for harpsichord or forte piano' (centre). The painting, by Josef Danhauser in 1840 (below), shows the young Franz Liszt playing for the writer George Sand (seated), Countess Marie d'Agoult (foreground), and Liszt's fellow composers Berlioz, Paganini and Rossini.

In the long run, two models prevailed. One was English, and it followed on from the work of Silbermann, imported across the Channel by his pupil Johannes Zempe. The pianos produced by John Broadwood and others were for the most part massive and rectangular, permitting greater string tension and so creating a larger, more robust sound. Such instruments were the ancestors of today's pianos. There was also a vogue for square pianos (above), promoted at a much publicised concert given by Johann Christian Bach, son of Johann Sebastian, in London in 1768.

In Austria a different tradition developed under the influence of another of Silbermann's pupils, Johann Andreas Stein. His pianos were physically less heavy than their English counterparts, and also required a lighter touch from the pianist, producing a warm, lilting timbre. Like harpsichords, they had several registers, activated by pedals or knee levers, that could modify the sound of the chords (for example by raising the dampers or by interposing a strip of felt). These German and Viennese instruments were the ones used by Mozart, Haydn and the young Beethoven.

A generation of musician-composers were soon putting the new instruments to the test with virtuoso pieces that stretched their capacities to the limit. This was the golden age of the piano sonata and the concerto, a form pioneered by Johann Christian Bach and raised to new heights by his successors.

From classicism to romanticism

Piano-making flourished in the early 19th century, notably in Germany by Johann Andreas Stein, in Vienna by Nanette Streicher (Stein's daughter) and Anton Walter, and in France by Henri Pape. Some of the early Viennese pianos had keys in reverse colouring, with the naturals in black and accidentals in white. The tonal range came to stretch from the barely audible triple piano (*ppp*) to the deafening triple forte (*fff*). The internal action also became more efficient, permitting the rapid repetition of notes. The piano's range (the distance from the lowest to the highest pitch) increased from five octaves, to six and a half, and then to seven octaves by mid century. The use of pedals spread.

Piano-playing became more sensitive and delicate, encompassing greater contrasts of mood and tone and enabling performers to express a wide range of emotions. The composers of the Romantic era fully exploited these characteristics. Schubert, Schumann and, above all, Liszt and Chopin extended the instrument's repertoire, coming up with the

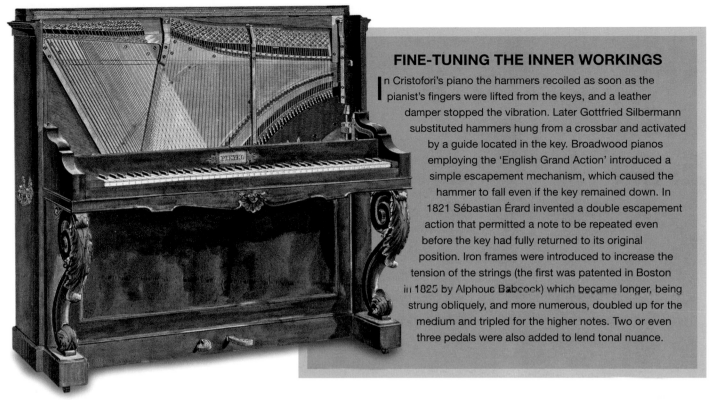

FINE-TUNING THE INNER WORKINGS

In Cristofori's piano the hammers recoiled as soon as the pianist's fingers were lifted from the keys, and a leather damper stopped the vibration. Later Gottfried Silbermann substituted hammers hung from a crossbar and activated by a guide located in the key. Broadwood pianos employing the 'English Grand Action' introduced a simple escapement mechanism, which caused the hammer to fall even if the key remained down. In 1821 Sébastian Érard invented a double escapement action that permitted a note to be repeated even before the key had fully returned to its original position. Iron frames were introduced to increase the tension of the strings (the first was patented in Boston in 1825 by Alphous Babcock) which became longer, being strung obliquely, and more numerous, doubled up for the medium and tripled for the higher notes. Two or even three pedals were also added to lend tonal nuance.

nocturne, the impromptu, the prelude and the *lied*, in which the piano accompanied the human voice. Their work ran the gamut from the most intimate musings to the full drama of the passions – so full in Liszt's case that on several occasions he broke piano strings during recitals. Steinway and Sons, founded in New York by a German immigrant in 1853, sought to address the problem by designing pianos for the most exacting virtuosos.

Modern pianos

At the start of the 20th century square pianos and the so-called 'giraffes' – a type devised in Vienna around the turn of the 19th century in which the sounding board curled vertically upward rather than stretching backward – went out of fashion, to be replaced by the familiar modern uprights and concert grands.

The most polyphonic of all instruments, the piano continued to be a vehicle for both solo and orchestral works, some reflective, some showily dramatic. It lent itself to styles as diverse as the impressionism of Debussy, Ravel's expressionism and the lyricism of Rachmaninov or Prokofiev. It was the vehicle of choice for ragtime and played a significant part in early jazz. Contemporary composers have used it to conduct experiments in sound; John Cage, for instance, composed pieces for 'prepared piano', the preparation consisting in putting nuts, bolts and pieces of rubber between

and among the strings. In recent decades the instrument has played a significant part in many kinds of popular music, from solo singer-songwriter albums to rock and soul.

From the 1960s on, there have been various technological spin-offs, among them the Rhodes and other electric pianos. Today, digital instruments compete with acoustic models; easily transported and never out of tune, they can be played using headphones and offer an eclectic range of sound effects.

Upright piano
A model with crossed strings (above left) made by Pleyel, Wolff and Co in 1880.

Youthful prodigy
Born in 1982, the Chinese pianist Lang Lang (below) is one of the latest in a long line of virtuoso piano performers to attract international admiration.

ACADEMIES AND SALONS
Science enters the Age of Enlightenment

From the 16th century on, first the Reformation and then the challenges to traditional teaching arising from the new scientific discoveries forced Western centres of learning to rethink their role. Academies, salons, learned correspondence and travel encouraged the spread of knowledge and the growth of a scholarly community across Europe.

University of Leiden
The Dutch physician Herman Boerhaave delivering a lecture at the oldest university in the Low Countries, founded in 1575 by William of Orange.

Anatomical theatre
Opened in 1584 in the Bo Palace, which had been home to the University of Padua since 1493, the hall had a ground-breaking design that would be a model for similar theatres built in later years.

Luther and Calvin were in agreement: the authority of the Scriptures extended only to matters of religion. In the lands where the Reformation took hold, scientific disciplines such as astronomy, mathematics, chemistry, medicine and the natural sciences were freed from the grip of theologians. The intellectual climate encouraged freedom of thought.

Linked to this movement and fed by the growing thirst for knowledge – whether for professional advantage or for the power that it could bring – was a drive to create new universities. This led to new foundations being established at Giessen in Hesse-Darmstadt in 1607 and at Kiel in Schleswig-Holstein in 1665. The spread of learning also breathed fresh life into existing institutions in the Protestant lands of northern Europe, England, Scotland and Sweden among them.

The Continent ended up divided between territories enjoying the educational freedom of the Reformation and those subject to the authority of Rome. Some universities founded in medieval times were still hamstrung by the sterile dispute between nominalists, who held that abstract qualities have no real existence, and realists, who insisted that they do. The curriculum here was still dominated by traditional disciplines such as rhetoric and scholastic philosophy.

In France there was a deep divide between traditional education, which was dominated by the Jesuits, and the free-thinking new spirit of the Scientific Revolution. At the start of the 17th century, the 40 colleges of the University of Paris were all but deserted.

New cauldrons of culture

Soon, however, academies started springing up, acting as alternative centres for the diffusion of knowledge. These were gathering points where intellectuals could freely exchange new ideas about the arts, archaeology or the sciences without worrying about the reception their views would receive. Scholars who otherwise rarely left their studies could compare findings with others similarly engaged, taking encouragement from their praise and learning from their criticisms.

There was nothing new about the institutions themselves: Florence's Platonic Academy, founded in 1442, had played host to Machiavelli, Pico della Mirandola, Politian and other Renaissance luminaries. By the late 17th century, however, academies were flourishing across Europe. Some specialised in particular areas, like the Academia Naturae

Oxford and Cambridge

England's two great academic rivals were founded respectively in the early 12th and early 13th centuries. For many years Oxford was considered strongest in the humanities, while Cambridge had the edge in the sciences. The books at left are in an Oxford library; below, the courtyard of Trinity College, Cambridge.

GUARDIANS OF FREE THOUGHT

By encouraging liberty of conscience and the separation of Church and State, the Reformation helped to free universities from the theologians' grip. The first Lutheran university was founded at Marburg in Germany by Philip I of Hesse in 1527, just ten years after Luther had raised the flag of revolt. Yet even in the Protestant lands the effects of the break were neither uniform nor sudden. When the reform movement reached the University of Basel, ideological disputes set the professors, who opposed it, against the local burghers, who were mostly in favour. The university's statutes were eventually revised, but it took a century to effect the change. At Oxford and Cambridge the curriculum evolved to emphasie Latin, Greek and mathematics in place of scholastic philosophy. Religion, politics and science were inextricably mixed, but scholarship fortunately managed to extricate itself from the disputes.

The French Academy of Sciences

The academy got underway relatively informally, meeting twice weekly in the royal library in the Rue Vivienne in central Paris. The painting below shows the academy's founder, France's finance minister Jean-Baptiste Colbert, presenting the academicians to King Louis XIV in 1667.

Curiosi, founded in 1662 in Leipzig, which was dedicated to the natural sciences, or the informal 'invisible college' that met in London and Oxford from 1645 on, prefiguring the establishment of the Royal Society in 1660. Others were multi-disciplinary, like the French Academy of Sciences, set up at the urging of Louis XIV's finance minister Colbert in 1666, whose members included the philosophers René Descartes and Blaise Pascal, the mathematician Pierre Gassendi, and even Sir Isaac Newton. In 1700 the Brandenburg Society of Sciences began proceedings in Berlin under the presidency of the philosopher Leibniz. A year later it was reborn as the Royal Prussian Academy of Sciences when its sponsor, Frederick III, became king. Twenty-five years later Catherine, the wife of Peter the Great of Russia, established the Imperial Academy of Sciences in St Petersburg, which would win enhanced international prestige half a century later under her illustrious namesake,

A MONK OF MANY PARTS

The French philosopher and monk Marin Marsenne (1588–1648) was one of the most influential actors in the scientific revolution of his day. Even though not himself a scientist in the strict sense of the word, he played a vital role as an organiser and supporter of those who were. Schooled in the same Jesuit-run college that trained René Descartes, he joined the Minim Friars and spent some years teaching before moving to Paris. He took up residence in a monastery there, making his base the hub of an active scientific and philosophic circle. Marsenne had eclectic interests: he served as a translator and publisher, maintaining correspondence with most of the leading intellectuals of his day, and he was a pioneer of musical theory, approaching acoustics from a scientific viewpoint in his *Traité de l'harmonie universelle* ('Treatise on Universal Harmony'), published in 1636.

Catherine the Great. These institutions played a significant role in democratising science, for their members held their discussions, and published their findings, in their native languages rather than in Latin, thereby increasing the audience for their work.

Despite the self-censorship forced on some of the academies as the price of royal patronage (see box below), the new institutions brought a breath of fresh air to intellectual life. Between the 16th and 19th centuries, no fewer than 2,200 academies were established in Italy alone. Starting out as unofficial coteries, the academies gradually found themselves transformed into state-supported institutions rivalling and sometimes even outstripping the universities.

The time of the scientist

Meanwhile, scientists came to enjoy an elite status and their work became a matter of general interest. Members of the academies travelled widely and exchanged ideas with fellow academicians in other countries. Following a pattern that had first become apparent in the Renaissance, they conducted correspondences that helped to create international networks. The Royal Society in London, whose motto was *Nullis in Verba* ('Take nobody's word for it'), not only strove for scientific objectivity but also for the free flow of scientific information across borders. The Society appointed fellows from other countries, including the Dutch scientist Christiaan Huygens, who subsequently acted as an adviser to Colbert when he was planning the French Academy of Sciences. Another Society member, the French mathematician Pierre Louis Maupertuis, went on to become the first president of the Berlin Academy of Sciences. The Swiss mathematicians Nicholas and Daniel Bernoulli were invited to join the Imperial Academy in St Petersburg, and so the list went on. In addition, most academies accepted correspondence members, who used the mail services to sent word of researches conducted in their native countries.

The spread of knowledge

The academies did not take in students and handed out no diplomas; instead, they published their members' research. Their influence was felt first in the court circles of the rulers who sponsored them, then more widely among the educated classes as a whole, spread by reviews and books written either in the local language or in French, which served at the time as the lingua franca of the literati.

One offshoot of the movement was a vogue for cabinets of curiosities – collections of objects related to the natural sciences, such as stuffed animals, fossils, minerals and dried plants. These private stockpiles were precursors of museums,

Founder's medal
A badge struck to mark the foundation of the French Academy of Sciences in 1667 (above).

TESTING THE LIMITS

From the 17th century on, members of learned bodies, such as London's Royal Society or the French Academy of Sciences, frequently enjoyed royal patronage. The honour was a mixed blessing, for it forced on the members a form of tacit self-censorship, particularly in the Latin countries of southern Europe. The celebrated Accademia dei Lincei ('Academy of the Lynx-eyed'), founded in Rome in 1603 by the naturalist Federico Cesi, welcomed Galileo as a member in 1611, but it was closed down by its own members after he was condemned by the Church; it only reopened in 1870. A similar body, the Accademia del Cimento ('Experimental Academy'), was set up in Florence in 1657 by Leopoldo de Medici, a disciple of Galileo who was brother to the grand duke of Tuscany. Like the Accademia dei Lincei it shut its doors after just a decade, but not before it had published important works in the fields of mathematics and physics, notably by Evangelista Torricelli. It, too, was re-established in the 19th century.

THE PARIS SALONS

Salon keeper
Many of the great French writers of the age, including Voltaire and Montesquieu, used to meet in the salon hosted by the Duchess of Maine (above) at the Chateau of Sceaux, outside Paris.

With their roots in court circles, the first salons came into existence in the course of the 17th century, started for the most part by women seeking to make their social mark. Mocked by the comic dramatist Molière as places where *précieuses* (affected women) competed to show off their exaggerated intellectual attainments, these gatherings established themselves first as sounding-boards for intellectuals and artists and then as focal points for the spread of Enlightenment ideas. The physician and mathematician d'Alembert, founder with Denis Diderot of France's famous *Encyclopédie*, was a regular guest at several salons. Benjamin Franklin attended the salon run by Mme du Deffand; Mme Necker's salon was popular with members of the Académie Française; that of Julie de Lespinasse was graced by the mathematician Condorcet. By that time the salons had come to serve as veritable parliaments of knowledge.

literally so in the case of the British Museum, which traces its roots back to the immense hoard accumulated by Sir Hans Sloane from the last years of the 17th century. Aristocratic salons, like the one hosted by the Duchess of Maine (1676–1753) from her chateau of Sceaux outside Paris, opened their doors to philosophers and other scholars. But there was a dark side to such popularity, for the prospect of a moment of glory in the salons attracted charlatans like Franz Anton Mesmer, apostle of 'animal magnetism', who claimed in the 1770s to be able to cure the sick by washing their feet in magnetised water.

Even so, science for the most part recognised its own, while the wider public swelled audiences for lecturers with something new to say. Mathematicians spelled out the law of probabilities in London gambling halls. In Edinburgh, students flocked to the university medical school, where courses were affordable and delivered in English rather than Latin, and where there was no obligation to graduate.

The spread of scientific knowledge did much to prepare the way for the ensuing Industrial Revolution. Experimentally verified knowledge provided the rulers and elites of nations with new ways of generating wealth.

By encouraging the spread of ideas, the salons and academies ultimately helped to change the world.

Conversation piece
A detail from a 1716 painting shows the French writers Fontenelle and La Motte with the mathematician Joseph Saurin at the salon of the Marquise de Tencin.

The tuning fork 1711

One enduring difficulty of ensemble playing in music is achieving perfect harmony. In adjusting the sound of their instruments, early musicians had to rely on their ears alone. Yet few, however talented, had perfect pitch. As a result, tuning instruments in medieval times was a fairly random process.

A reference point for musicians

Introduced in the 9th century, musical notation served to transcribe intervals but could not capture pitch. Two instruments of the same type would not necessarily produce exactly the same note. The correct tuning varied with locality, musical styles and periods: baroque-era instruments, for instance, were generally pitched lower than their modern equivalents.

By about 1690 people were using devices that emitted notes to serve as guides for groups of musicians jointly adjusting their instruments. Unfortunately these were not standardised, so the sound of the notes could vary from one ensemble to another. In 1711 an English musician named John Shore addressed this problem by introducing a metal fork equipped with a handle. When one of the tines of the fork was struck, both vibrated in harmony, emitting a pure musical tone.

Shore's invention is still in use today. Even so, the exact pitch of the note *la* and the other notes has changed in accordance with musical tastes, tending to move up the scale – a trend that ended up by creating difficulties for singers performing early works. Finally, in 1939, a conference held in London reached a consensus that the pitch of *la* should be set at 440 Hz. Since that time A440, or Concert A as it is known, has served as a standard for musical pitch recognised almost universally around the world.

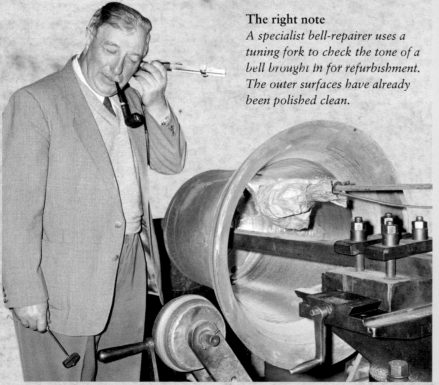

The right note
A specialist bell-repairer uses a tuning fork to check the tone of a bell brought in for refurbishment. The outer surfaces have already been polished clean.

RIVAL DEVICES

Pitch pipes, which are not very accurate, and purpose-built violin tuners have specialist applications but otherwise are not widely used. In contrast electronic tuners, which allow the user to adjust pitch very precisely, are becoming more and more common. The E-tuner (above) was devised by Laurence Equilbey, director of the French vocal ensemble Accentus. Used in concert for the first time in 2008, it comes with an earpiece that enables singers to access any note, frequency or micro-interval (an interval smaller than a semitone), opening up new possibilities for contemporary vocal music.

The art of measuring probabilities

Born of the calculation of probabilities and the law of large numbers, statistics is a mathematical discipline with important applications in many fields, from the social sciences and psychology to astronomy, physics and economics. It is fundamental to opinion polling and to medical epidemiology.

Toss a coin in the air. It has two sides, heads and tails, so there is a one-in-two chance that, say, tails will come up. The first throw turns up heads, then the second, third and fourth. Out of luck? Relax. The more often you throw the coin, the more the results will reflect the initial 50-50 odds, producing heads and tails in equal numbers. So says the law of large numbers: the likelihood of a phenomenon of a given probability occurring will tend towards the mean the more one multiplies the number of attempts. The first person to formulate the rule was the Swiss mathematician Jacques Bernoulli (1654–1705). Building on earlier work by Pierre de Fermat and Blaise Pascal, he paved the way for the theory of probabilities in his *Ars Conjectandi* ('The Art of Conjecture'), published posthumously in 1713, which laid the foundations of the science of statistics.

Laplace's demon and Gauss's curve

Several scientists subsequently developed Bernoulli's ideas. The French mathematician Pierre-Simon Laplace (1749–1827) defined a theory of errors, the name given to the deviation of a sample from the mean; so, in tossing a coin, the probability of throwing 10 heads in a row is less than the probability of throwing equal numbers of heads and tails. A convinced determinist (see panel below), Laplace famously believed that an intellect apprised of all the facts about the universe at any given time, from the largest heavenly bodies to the tiniest atoms, would be able to

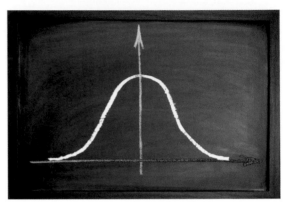

The Gauss curve
The bell curve represents the normal distribution pf data around a mean or average, represented by the peak of the curve. The variable factors fall away steeply on either side.

predict the future as clearly as it could see the past; later commentators would name this hypothetical brain 'Laplace's demon'.

The German mathematician Carl Friedrich Gauss (1777–1855) bequeathed to posterity the Gaussian function, a bell-shaped curve representing the different probabilities of any given phenomenon occurring. He showed the need to establish limits in the mathematical progression toward infinity, hugely advancing the calculation of probabilities.

Defining the average man

The word 'statistics' comes from the German *Statistik*, coined by the economist Gottfried Achenwall in 1746. Belgium's Adolphe Quételet (1796–1874) had the idea of applying statistical techniques to the social sciences. Astronomers, he noted, used statistical formulae to control the margin of error in their calculations. He was also aware that the bell curve could be applied to other kinds of measurements, such as the height of conscripts in a regiment. Putting the two facts together, he came up with a new concept: the average man. Even if every individual on Earth is different from all the others, people en masse nonetheless have between them a joint

A SUBJECTIVE PHILOSOPHICAL EXERCISE

From its early days, statistics contained the germ of an ideological conflict that was to rage through the 20th century. This set the deterministic view of nature, which can be traced from Laplace to Einstein, against a non-determinist conception that passed from Bernoulli to Heisenberg. While Einstein claimed that God 'does not throw dice', Heisenberg's uncertainty principle held that it was impossible even to predict the trajectory of a particle with complete accuracy.

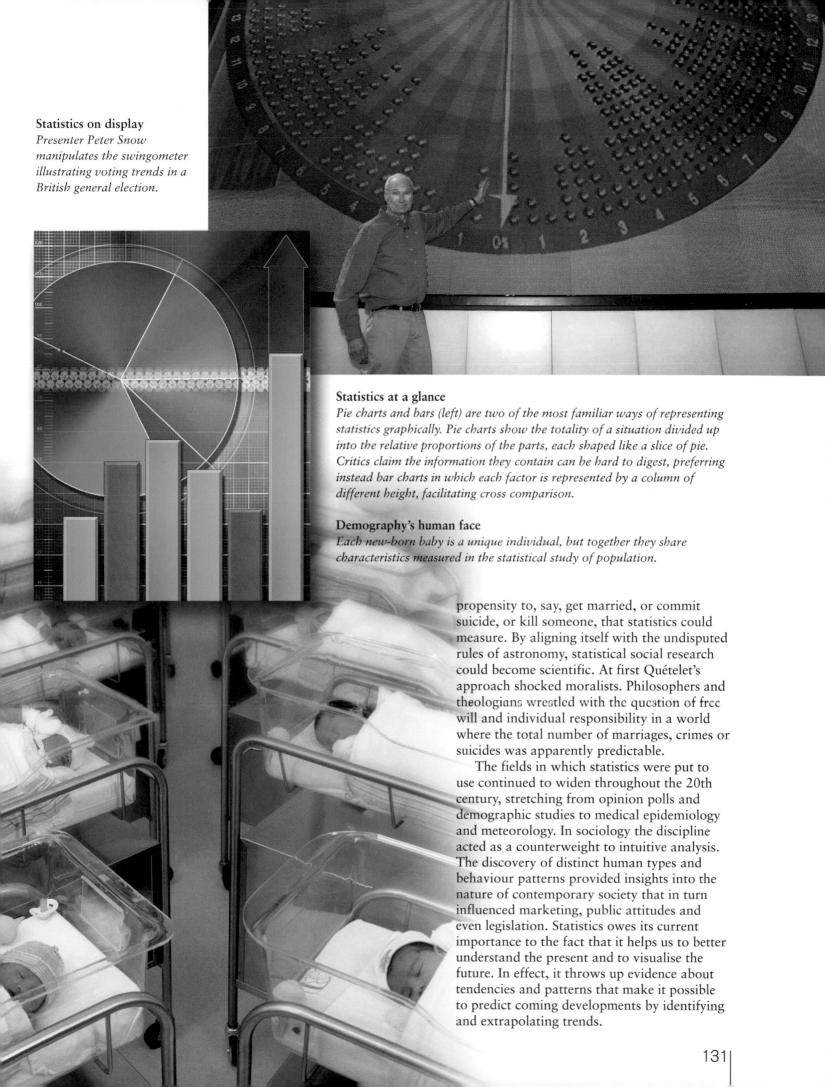

Statistics on display
Presenter Peter Snow manipulates the swingometer illustrating voting trends in a British general election.

Statistics at a glance
Pie charts and bars (left) are two of the most familiar ways of representing statistics graphically. Pie charts show the totality of a situation divided up into the relative proportions of the parts, each shaped like a slice of pie. Critics claim the information they contain can be hard to digest, preferring instead bar charts in which each factor is represented by a column of different height, facilitating cross comparison.

Demography's human face
Each new-born baby is a unique individual, but together they share characteristics measured in the statistical study of population.

propensity to, say, get married, or commit suicide, or kill someone, that statistics could measure. By aligning itself with the undisputed rules of astronomy, statistical social research could become scientific. At first Quételet's approach shocked moralists. Philosophers and theologians wrestled with the question of free will and individual responsibility in a world where the total number of marriages, crimes or suicides was apparently predictable.

The fields in which statistics were put to use continued to widen throughout the 20th century, stretching from opinion polls and demographic studies to medical epidemiology and meteorology. In sociology the discipline acted as a counterweight to intuitive analysis. The discovery of distinct human types and behaviour patterns provided insights into the nature of contemporary society that in turn influenced marketing, public attitudes and even legislation. Statistics owes its current importance to the fact that it helps us to better understand the present and to visualise the future. In effect, it throws up evidence about tendencies and patterns that make it possible to predict coming developments by identifying and extrapolating trends.

Eau de Cologne 1709

4711
Tosca
Eau de Cologne
Parfum

The use of perfume goes back to the earliest times. Composed of animal substances like musk, civet or ambergris, or of vegetable matter such as citrus, spices and certain flowers, traditional perfumes tended to be greasy. The idea of diluting the fragrance with a solvent such as alcohol first emerged at some point in the late 17th or early 18th centuries. Giovanni Paolo Feminis is

Fragrant delights
A 1950s advertising poster for Tosca, an eau de Cologne made by the 4711 company that set up in Cologne at the start of the 19th century.

sometimes credited with having created an *aqua mirabilis* ('wonder water') in 1695, after which the habit of referring to perfumes as *eaux* ('waters') was born. Later they would be known as *eaux de toilette*.

In 1708 the scent-maker Johann Maria Farina (1685–1766) created in his home town of Santa Maria Maggiore in northern Italy a fragrance so exquisite that he claimed in a letter to his brother: 'My perfume calls to mind a beautiful spring morning after the rain. It is made up of oranges, lemons, grapefruit and bergamot, of the flowers and fruits of my native land.' In 1709 Farina settled in the German city of Cologne where he began to sell his fragrance, which he named eau de Cologne in honour of the city; the French name reflected the international role of the language in trade and fashion. Eau de Cologne quickly conquered the courts of Europe, and has remained popular to this day.

Central heating 1716

In 1716 a Swede named Martin Triewald had the idea of heating a greenhouse in Newcastle-on-Tyne by means of a system that involved circulating hot water through concealed channels. Sixty-one years later a French architect, Jean Simon Bonnemain, adapted the principle to incubate eggs. At first this new type of heating using a circulating heating agent – warm air and steam were used as well as water – was reserved for industrial and agricultural use in buildings where the temperature needed to be kept high.

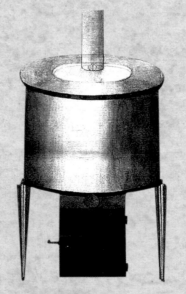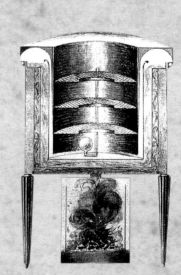

Bonnemain's incubator
The original system was installed on a large poultry farm that supplied the French court at Versailles as well as several Paris markets.

Turning the heat up

Human comfort first came into the picture in the 19th century, initially in public buildings such as churches, hospitals and prisons. Along with air conditioning and other technological spin-offs, central heating only put in an

appearance in private homes much later – almost twenty centuries after the invention of the hypocaust, an ancestor of the system that rich Roman families were employing as early as the 1st century BC.

4-colour printing c1719

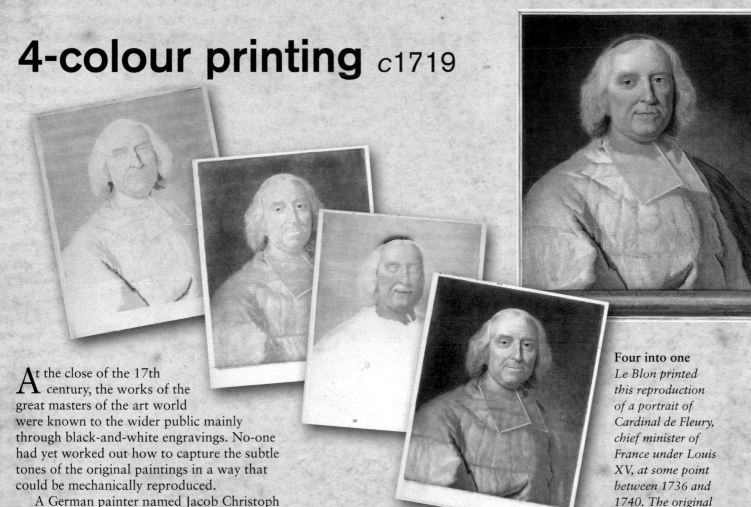

Four into one
Le Blon printed this reproduction of a portrait of Cardinal de Fleury, chief minister of France under Louis XV, at some point between 1736 and 1740. The original painting was by Hyacinthe Rigaud.

At the close of the 17th century, the works of the great masters of the art world were known to the wider public mainly through black-and-white engravings. No-one had yet worked out how to capture the subtle tones of the original paintings in a way that could be mechanically reproduced.

A German painter named Jacob Christoph Le Blon set out to address the problem. Putting Newton's researches on the spectrum into practice, he deconstructed each shade of the paintings into blue, yellow and red. In his system green was a mixture of blue and yellow, violet of red and blue. Le Blon then developed a range of inks that blended into one another rather than clashing. Finally, he had the idea of utilising the dark-to-light printing technique used for mezzotints, invented by the German Ludwig von Siegen in 1642. Combining the three ideas, he developed four-colour printing in the years from 1719.

The whole gamut of tones

To reproduce a painting, Le Blon prepared four separate copper plates, three for the primary colours of red, blue and yellow, with the fourth plate black. Guides on the plates ensured that successive impressions were perfectly aligned with one another. Each plate was individually subjected to the mezzotint process (see box, right); the blue plate, for example, was only marked in places where the colour blue was to appear. By superimposing the four impressions on a single plate, Le Blon recaptured all the tones of the original painting.

Failures and successes

Although Le Blon was granted exclusive rights to print multiple copies of paintings and coloured drawings, his process proved too time-consuming to be economically viable, and it fell out of use after his death in 1741. It came back into favour almost a century later in 1837, when it was adapted to the lithographic process by a printer from Alsace named Godefroy Engelmann, marking the start of chromolithography. In the 20th century the four-colour CMYK system – using Cyan (blue), Magenta (red), Yellow and Black (K) – finally came into its own with the development of offset printing.

THE MEZZOTINT PROCESS

The process involves roughening the surface of a copper plate with an instrument known as a rocker, marking it with a grid of tiny pits that retain ink, thereby creating deep tones. The surface can then be polished to smooth out the pits in areas where the image demands, so allowing for wide gradations of each colour.

Locating latitude at sea

Hadley's octant
An instrument made in the early 19th century to John Hadley's original design (below) by the London-based instrument-makers Spencer, Browning and Rust.

The 18th century was an era of great voyages of discovery, but navigation at sea remained a perilous enterprise. By developing the marine sextant, John Hadley introduced some welcome astronomical precision into the calculation of latitude.

In the years after 1600 maritime traffic was busier than it had ever been thanks to the discovery of the Americas, growing international commerce and the establishment of the first European colonies. More than 200,000 ships regularly travelled the world's sea-lanes, but as the numbers increased, so too did the frequency of shipwrecks caused by navigational errors, strewing the oceans with sunken vessels and drowned sailors. On November 2, 1707, four Royal Navy ships ran aground on rocks off the Scilly Isles with the loss of Sir Cloudesley Shovell, rear-admiral of the fleet, and some 1,400 men. In response Parliament, with the encouragement of the Royal Society, passed the Longitude Act of 1714, offering a prize of up to £20,000 for anyone who could devise an accurate method of fixing longitude at sea. One of the people who addressed themselves to solving this 'longitude problem' was the Society's own vice-president, a keen astronomer and skilled

instrument maker named John Hadley. He was a wealthy man and his scientific researches were driven more by curiosity than the hope of financial reward. He could hardly have guessed at the time that his work would end up by resolving an entirely different question – that of latitude.

Navigating by the stars

On-board navigation demands first and foremost a clear idea of where the ship is on the ocean's surface at any given time. The north–south position is defined in terms of latitude, the east–west location by longitude. Ever since antiquity, sailors had calculated latitude by taking readings from the Pole Star using an astrolabe. They would take successive readings for the horizon, for their own position and for the star itself. If the star appeared to be rising in the sky, that meant that the ship was heading north; if declining, then the direction was southward.

Such rough-and-ready methods were hardly adequate for the needs of 17th-century fleets, not least because south of the Equator there was no Pole Star to provide any bearings at all. And wherever they were, when the sky was overcast, mariners were forced back on the simple compass – known in the West since the 15th century – to work out a direction to steer. With the instruments available at the time and the additional margin of error brought about by the pitching of the ship, it was not

In peril on the sea
Before the invention of the sextant, navigation was a hazardous business at the best of times, let alone in storms like the one that sank HMS Ramillies (left) off the coast of Newfoundland during the American War of Independence.

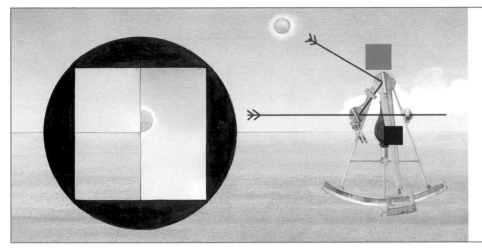

uncommon for a captain to be 4° or 5° out in his calculations – an error that equated to between 250 and 300 miles.

A simple and ingenious principle

In 1731 Hadley presented to the Royal Society a description of a new instrument for calculating latitude. The principle was straightforward, but ingenious: it required two mirrors to bring the images of the heavenly body being viewed and of the horizon into alignment via a split-screen effect. Unlike the ancient astrolabe, this new instrument allowed observers to see both objects whose position they wished to establish at one and the same time. It featured a telescopic sight equipped with a graduated micrometer, through which the navigator could observe the horizon and the image reflected by the two mirrors. Solidly built, the instrument was almost

unaffected by the rolling of the ship. It reduced inaccuracies in measurements of latitude to just three seconds (less than 0.001 of a degree), equating to about half a mile.

Strictly speaking, the instrument that Hadley invented was an octant, with an arc extending over 45°, or one-eighth of a full circle. Captain John Campbell followed Hadley's design to produce an instrument with a 60° arc (one-sixth of a circle) in 1757, thereby creating the modern sextant.

A lingering problem

Thanks to the sextant the problem of latitude was solved, but that of longitude remained. Astronomers produced ephemerides – astronomical tables giving the exact position of the heavenly bodies for given times and locations – but observing the stars from on board ship presented the same difficulties that it always had, making measurements as imprecise as ever. A solution had to wait on the development of highly accurate chronometers that could link celestial coordinates to fixed terrestrial ones via an international time standard, permitting the user to calculate the ship's east–west position with reasonable precision. The prize offered by Parliament was never awarded. A smaller sum was eventually given in compensation to the man who developed that super-accurate timepiece, the Yorkshire clockmaker John Harrison – but that, as they say, is another story.

Sextant sightings
Sextants (right) had half-horizon mirrors that split the field of view in two (top) to show the object observed (here the Sun) in line with the horizon. The user could then read off the angle in degrees.

Marine chronometer
John Harrison's pioneering device (the model above dates from 1770) was developed in response to the need for a precisely accurate timepiece to calculate longitude – the distance east or west from the Greenwich meridian.

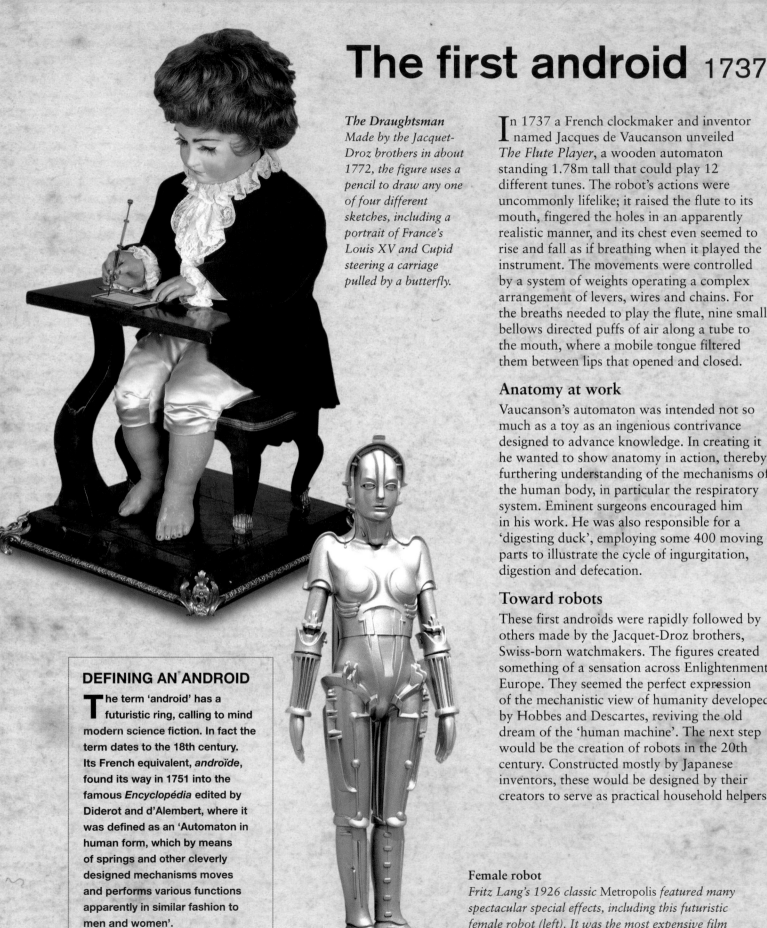

The first android 1737

*The Draughtsman
Made by the Jacquet-Droz brothers in about 1772, the figure uses a pencil to draw any one of four different sketches, including a portrait of France's Louis XV and Cupid steering a carriage pulled by a butterfly.*

In 1737 a French clockmaker and inventor named Jacques de Vaucanson unveiled *The Flute Player*, a wooden automaton standing 1.78m tall that could play 12 different tunes. The robot's actions were uncommonly lifelike; it raised the flute to its mouth, fingered the holes in an apparently realistic manner, and its chest even seemed to rise and fall as if breathing when it played the instrument. The movements were controlled by a system of weights operating a complex arrangement of levers, wires and chains. For the breaths needed to play the flute, nine small bellows directed puffs of air along a tube to the mouth, where a mobile tongue filtered them between lips that opened and closed.

Anatomy at work

Vaucanson's automaton was intended not so much as a toy as an ingenious contrivance designed to advance knowledge. In creating it he wanted to show anatomy in action, thereby furthering understanding of the mechanisms of the human body, in particular the respiratory system. Eminent surgeons encouraged him in his work. He was also responsible for a 'digesting duck', employing some 400 moving parts to illustrate the cycle of ingurgitation, digestion and defecation.

Toward robots

These first androids were rapidly followed by others made by the Jacquet-Droz brothers, Swiss-born watchmakers. The figures created something of a sensation across Enlightenment Europe. They seemed the perfect expression of the mechanistic view of humanity developed by Hobbes and Descartes, reviving the old dream of the 'human machine'. The next step would be the creation of robots in the 20th century. Constructed mostly by Japanese inventors, these would be designed by their creators to serve as practical household helpers.

DEFINING AN ANDROID

The term 'android' has a futuristic ring, calling to mind modern science fiction. In fact the term dates to the 18th century. Its French equivalent, *androïde*, found its way in 1751 into the famous *Encyclopédia* edited by Diderot and d'Alembert, where it was defined as an 'Automaton in human form, which by means of springs and other cleverly designed mechanisms moves and performs various functions apparently in similar fashion to men and women'.

*Female robot
Fritz Lang's 1926 classic* Metropolis *featured many spectacular special effects, including this futuristic female robot (left). It was the most expensive film to be made in Germany in the silent era.*

Sparkling water 1741

Esteemed since ancient times for their medicinal qualities, naturally sparkling mineral waters were much studied from the 16th century on by those seeking to discover the secrets of their effervescence. In 1741 William Brownrigg, a Cumberland physician, became the first person to produce the sparkling effect artificially by introducing salts such as bicarbonate of soda into ordinary spring water. He also managed to demonstrate the presence of carbonic acid in the water. The celebrated chemist Joseph Priestley was the first person to come up with a method of injecting carbon dioxide into water by dissolving it under pressure, thereby opening the way to the future bottled-drink industry.

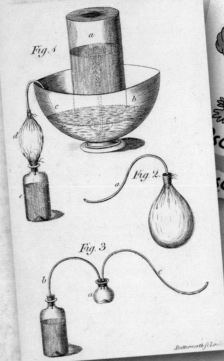

Adding bubbles
An illustration of Priestley's method for carbonating water (left). Sparkling drinks are now ubiquitous: the cartoon above is a 1956 poster advertising a French variety.

THE LEYDEN JAR

Across Europe other scientists set out to repeat von Kleist's experiment. In January 1746, a Dutch scientist named Pieter van Musschenbroek reported his discomfort in a letter to the French Academy of Sciences: 'My whole body was jolted as if by lightning … I would not expose myself again to such a shock for the crown of France.' Van Musschenbroek was a professor at the University of Leiden, so the first capacitors became known as Leyden jars (below).

The capacitor 1745

At the start of the 18th century electricity was little understood, but research conducted on static electricity meant that the subject was very much in people's minds. The experiments excited so much attention that public demonstrations were organised in fairs and salons. That was how the German clergyman-physicist Ewald Jürgen von Kleist happened, in October 1745, to plunge a metallic rod charged from a friction machine into a foil-lined bottle of water he was holding in one hand. In doing so he forgot to follow the advice of one of his predecessors, who had recommended that the receptacle should be well insulated, and he received a sharp shock when his hand touched the bottle. In effect he had created a capacitor – a device capable of storing electricity that is composed of two conductors separated by an insulator. In von Kleist's experiment, his own hand and the water were the conductors, while the glass of the bottle was the dielectric or insulator storing the charge.

SUGAR BEET – 1745
A root to rival sugar cane

Long considered a luxury, sugar became even more so when a blockade imposed by Napoleon stopped sugar cane from British colonies reaching continental Europe. But necessity bred invention: a viable way of processing sugar beet was found and when the new sugar was introduced, no-one noticed the difference. Sugar soon became an everyday commodity.

Sugar beet and fodder beet
These lithograph illustrations of different types of beet (right) are from the Album Bernary, *published by the German botanist Ernst Bernary in 1876.*

Preparing syrup
A 15th-century illustration of the production of syrup from sugar cane.

Andreas Sigismund Marggraf, an apothecary and chemist, was an eminent member of the Berlin Academy of Sciences when, in 1745, he published the work that introduced sugar beet to the world. It was titled *Chemical Experiments Made with a View to Extracting Sugar from Divers Plants that Grow in our Lands.*

People had in fact known since antiquity that there was sugar in beets, but in Marggraf's day the plant served primarily as a laxative. The idea of using it as a source of table sugar had seemed a ridiculous one.

Ever since the 8th century, the West and the Orient alike had got their sugar not from beet but from sugar cane, grown mainly in Spain and the south of France. Christopher Columbus took it to the New World in 1493, and its successful introduction to Santo Domingo (the future Dominican Republic) encouraged colonists to create sugar cane plantations across the Caribbean. In 17th-century Europe sugar was a sought-after luxury; sugar crystals were spoken of in the same breath as precious stones, and sometimes were even listed in princesses' dowries.

Marggraf ground up fodder beet, extracting a juice that he then filtered and concentrated by heating. The crystalline residue tasted like cane sugar, but the process was not viable economically, delivering only a few grams per kilo processed.

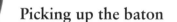

Picking up the baton

Half a century went by and the only person to show much interest in Marggraf's work was one of his pupils, Franz Karl Achard. Having worked out the variety of beet that would produce the best yield, he got official backing from the King of Prussia, Frederick William III, to establish a production plant in Silesia. Marggraf's idea was at last beginning to bear fruit, but Marggraf himself did not know it, having died in 1782. Meanwhile the growing taste for chocolate, coffee and tea had substantially increased the demand for sugar.

Achard's initiative attracted attention across Europe, and nowhere more so than in Britain. Fearing that the new source might kill off the sugar cane industry in the colonies, British merchants offered Achard 200,000 *taler* to declare his experiment a failure, but he turned them down. Even so, beet proved difficult to commercialise; it was more complex to process than cane, and at first produced a lower-quality product that proved uncompetitive in the market-place. The French Academy of Sciences expressed reservations as to its future. But then history intervened.

Filling a need

In 1806 France lost her Caribbean colonies in the course of the Napoleonic wars. Napoleon responded by introducing the Continental System, which prohibited the areas of Europe under his control from importing goods from Britain or her colonies. The results were dramatic; although Britain continued to import cane sugar from the colonies, sugar imports in France collapsed from 25,000 to 2,000 tonnes.

Demand did not fall, however, so suddenly there was an urgent need for new sources of supply. In response, the Emperor's leading pharmacist, Nicolas Deyeux, set to

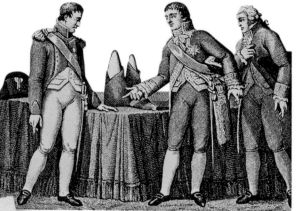

Sweet success
France's Minister of the Interior presents beet sugar to the Emperor Napoleon (left).

work to improve Achard's techniques. In 1810 he told the Academy of Sciences that large-scale production would be needed to support a viable industry. A philanthropic banker, Benjamin Delessert, responded to the appeal. Delessert had made a fortune as co-founder of France's first savings bank and was also known for founding soup kitchens for the poor. He was looking for a new use for some production facilities that had been running well below capacity since the blockade came into force. In 1811 he found a way of mechanising Achard's process using steam-operated machinery. His success was enough to persuade Napoleon to set up a chain of 40 refineries across France. In 1838 the industry crossed the Atlantic when a first refinery was established in the USA. Sugar beet had finally moved onto the international stage.

Sugar and strife
A 19th-century German cartoon (left) shows sugar cane at war with sugar beet.

Sugar beet production
The illustration below shows sugar beet being processed for sugar in the early 19th century.

FROM BEET TO SUGAR

Washed and sliced, the beets are fed into a cylinder where warm water, flowing in the opposite direction, is sweetened by the sugar contained in the roots. The juice is cleaned, filtered and heated, raising the sugar content from 13 to 70 per cent. The resultant juice is poured into boilers to form a paste called massecuite in which the sugar crystals are held in suspension. Centrifuges then separate the crystals from the remaining syrup. Dried in warm air, the sugar is ready for consumption as soon as it has cooled.

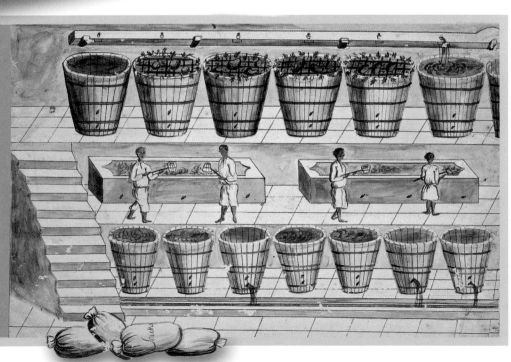

COSMOPOLITAN AMSTERDAM

A 17th-century metropolis

In the 17th century the city of Amsterdam was a shining beacon, its economic dynamism equalled by its intellectual vitality. The ships that thronged the city's port made it the market place of the world, while the unusual tolerance and humanism of the authorities made this hub of international commerce a haven for immigrants. The cultural mix served to enhance the city's standing as a centre for painters, publishers and printers.

Art meets trade
Rembrandt's
painting The
Drapers' Guild
(below) shows
the Guild's members
– prominent
representatives of
the city's commercial
elite – checking their
accounts in 1662.

In the port a sea wind stirred a forest of ships' masts. The vessels they belonged to had sailed there from all over the world. The warehouses lining the quays housed a cornucopia of goods: ores from Sweden, timber from Norway, oil and fruits from Spain, woollen goods from England, sugar, spices, tobacco … Shops selling silks, carpets and jewellery attracted the wealthy to the Dam, the main square where the town hall stood. Prosperous burghers' houses looked out over tree-shaded canals. Such was Amsterdam in the mid 17th century.

A radical transformation

The extraordinary prosperity of this 'pearl of the world's cities' was in marked contrast to the economic somnolence of its main European rivals. The days were long gone when the settlement

City plan
A map of
Amsterdam in
1662 (above)

Abjuration, cutting the colonial ties that had formerly bound the nation to Spain and declaring independence. In the wake of the split, the United Provinces of the Netherlands became a federal republic that quickly made itself a leading economic power, in effect inventing the mechanisms of the free market. In this northern corner of Europe, the ports welcomed foreign trade, and taxes on goods passing through them were kept low.

By the start of the 17th century, two thirds of the ships that sailed into the Baltic Sea flew the Dutch flag. Farther afield, the Dutch East India Company exerted a quasi-monopolistic grip on the spice trade with Indonesia and Sri Lanka, as well as over the traffic in Far Eastern porcelain and textiles from India. There was also a Dutch West India Company that engaged in the sinister but lucrative African slave trade, bringing back sugar from Caribbean plantations on return journeys.

Amsterdam was the commercial hub of the young republic. It built its success on the demise of its rival Antwerp, which was retaken by the Spanish in 1585 and then cut off from the Scheldt estuary by the United Provinces fleet. Suddenly the extra 75 miles of sometimes

that grew up at the mouth of the Amstel River in the early 13th century had been little more than a fishing port serving the herring trade. But thanks partly to the prosperity brought by the fish, partly to the sheer industriousness of the inhabitants, the town prospered, its population growing from 30,000 in 1585 to 150,000 in 1650 to more than 200,000 in 1675. By that time Amsterdam, once ringed by lakes and threatened by the sea, had been transformed. The citizens had undertaken major drainage works. Four great canals in concentric semi-circles, linked by a network of smaller waterways and side streets, had pushed the city limits back well beyond their former bounds.

The Dutch miracle

Politics played a part in Amsterdam's rise. On July 2, 1581, the Dutch parliament, known as the Estates-General, passed the Act of

AN OASIS IN THE HEART OF THE CITY

Founded by the city fathers in 1638, Amsterdam's Hortus Botanicus is one of the world's oldest botanical gardens. In its early years it sheltered a rich collection of medicinal plants for the use of the city's doctors and

apothecaries. It was moved to the Plantage district in 1682, where its collections were later enriched by species brought back by Dutch East India Company merchants. Today it houses some 4,000 plants drawn from all over the world.

The port of Amsterdam
The city joined the Hanseatic League in 1368, and soon became a vital commercial centre, serving as a link between the North Sea and Baltic trade routes. The cityscape (top centre) was painted by Abraham Storck.

Company warehouses
The Dutch East India Company was founded in 1602, when the Dutch Estates-General (parliament) brought several existing smaller bodies together in a single institution.

amounts of cash. The city's stock exchange was a vital organ of international commerce, providing a place for buyers and sellers to do business without the petty inconveniences attendant on conducting affairs in warehouses or the street. The dockyards were hives of activity, and the city's tobacco and soap factories, sugar refineries and diamond-cutting workshops were kept busy transforming the raw materials that were carried in on each tide. The general level of activity was so intense that France's great finance minister Jean-Baptiste Colbert was moved to claim that the average Dutchman did more work in a day than his own compatriots did in a week.

A breath of liberty

Above all, Amsterdam willingly opened its gates to the flood of refugees who found their way to the city. Antwerp merchants and Protestant Huguenots driven out of France for their religious beliefs sought refuge there, putting their talents and their commercial know-how at the service of their new homeland. The philosopher René Descartes, who lived on the banks of the Amstel from 1629 to 1635, hit the mark when he said of the city, 'Where else can a man enjoy such complete liberty, go to sleep each night with less apprehension, or find more traces of our forebears' innocence?'

dangerous waterways that separated the two ports was no longer a deterrent to shipping. From all over the world a steady flow of heavily laden vessels converged on Amsterdam's docks.

The city burghers took steps to welcome them. The Bank of Amsterdam was established in 1609 to regulate foreign currency dealings and the money supply, issuing bankers' orders that did away with the need for merchants to carry large

Stock Exchange
The Amsterdam Bourse (top) was established on the initiative of the Dutch East India Company in 1602, making it the world's oldest.

The *Amsterdam Gazette*
A French-language weekly (above) published in the city from 1691 to 1796.

Monkey business
The painter Jan Brueghel the Younger (1601–78) caricatured tulipmania in this 1640 canvas (right), portraying buyers and sellers of the bulbs as apes.

TULIPMANIA HITS AMSTERDAM

In the late 16th century a Flemish botanist named Charles de l'Écluse introduced tulips to Holland from Turkey. The Dutch were entranced by the flower, which horticulturalists produced in many different varieties. Soon speculators, lured by the prospect of quick profits, were selling everything they possessed to buy the bulbs. Between 1633 and 1636, when tulipmania was at its peak, the price of bulbs rose by up to 5,900 per cent, and the first of the great speculative bubbles that have marked the development of Western society went into overdrive. By the early weeks of 1637 bulbs were selling for up to 5,200 guineas, the price of a well-appointed merchant's house. The bubble burst in February 1637, ruining many people and almost bringing down the entire Dutch economy.

By the end of the 17th century Amsterdam had street lighting that allowed pedestrians to make their way around the city at night in relative safety, while fire pumps reduced the risks of conflagrations. It owed the fire pumps to the painter Jan van der Heyden, a specialist in architectural studies who was also Amsterdam's chief fire officer.

An intellectual ferment

Unsurprisingly, the spirit of liberty that distinguished Europe's most cosmopolitan city, combined with the relative calm and security it offered, made it a paradise for publishing houses and bookshops. Dozens were already in operation by the 1670s. While much of the rest of Europe lived in the shadow of royal absolutism, the United Provinces had no systematic censorship, and books were produced there that would never have seen the light of day in most parts of the Continent. The press, too, was in rude health, and there was strong public demand for gazettes that appeared once, twice or three times a week, their pages full of classified advertisements.

Even though Amsterdam was a crossroads for Europe's intelligentsia, it did not have a university of its own, like Leiden had. Instead it claimed to shelter the most important of the so-called 'illustrious schools' – elite institutions that set out to provide a superior education for their pupils.

The city was also home to a galaxy of gifted artists, the greatest of them being Rembrandt van Rijn. Rembrandt came from Leiden, his hometown, to serve an apprenticeship in Amsterdam when he was 18 years old, and returned to live there in 1631, seven years later. He was hugely successful, but he also lived beyond his means and in 1656 was forced to sell most of his belongings in order to avoid bankruptcy. He died alone in 1669.

Rembrandt's decline in a sense foretokened that of the city itself. Amsterdam and the whole province of Holland where it lay were overshadowed in the 18th century by Britain's emergence as a dominant mercantile power. The city's fortunes declined as this new rival cast its net over the commercial dealings of the continent and eventually the world.

The Night Watch
Painted in 1642, Rembrandt's masterpiece (top) portrays one of the companies of civic militiamen who served as city guards in the days before the creation of a permanent police force.

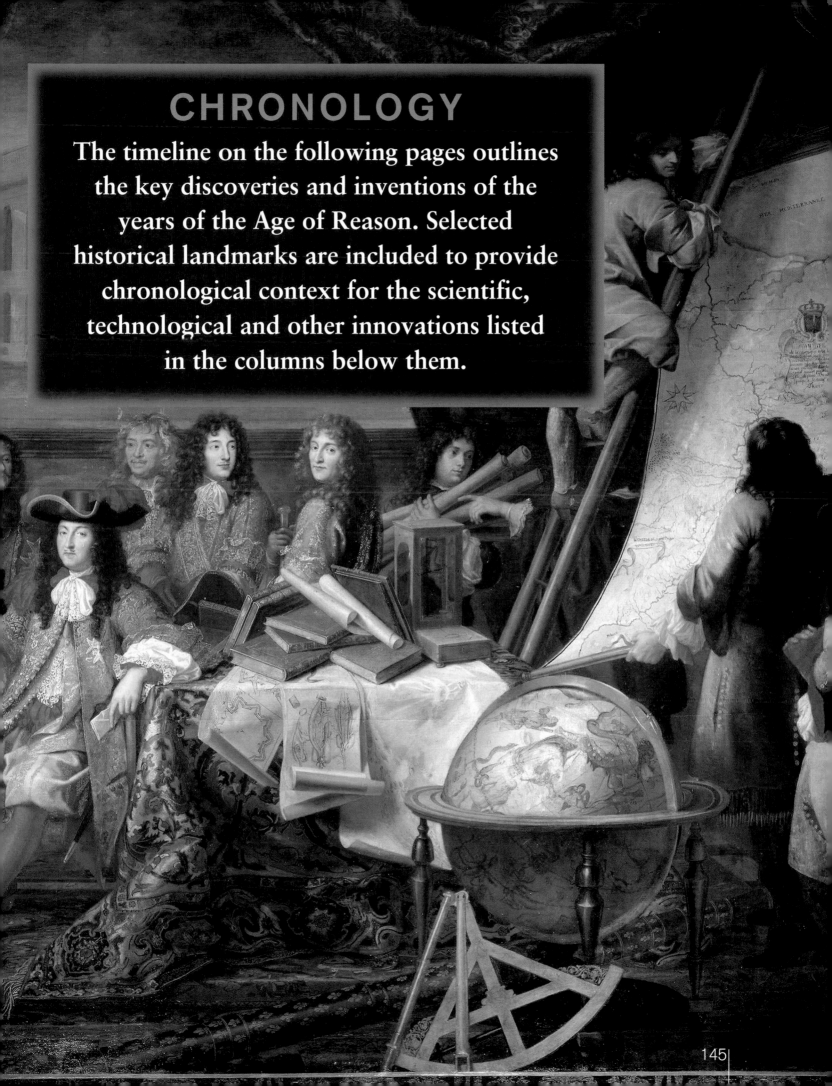

CHRONOLOGY

The timeline on the following pages outlines the key discoveries and inventions of the years of the Age of Reason. Selected historical landmarks are included to provide chronological context for the scientific, technological and other innovations listed in the columns below them.

1590

EVENTS

- William Shakespeare writes his history plays
- Henri IV ends the French wars of religion (1598)
- Shah Abbas makes Isfahan the capital of Persia (1598)

INVENTIONS

- Galileo formulates the law of falling objects (1590)

- France's first botanical garden is established by the University of Montpellier (1593)

- Johannes Kepler publishes his first major astronomical work, the *Mysterium Cosmographicum* (1596)

- Sir John Harrington invents the flush toilet (1596)

1600

- Giordano Bruno, Italian mathematician and astronomer, is burned at the stake for heresy in Rome (1600)
- The Dutch East India Company is founded (1602)
- Death of Queen Elizabeth I of England (1603)
- Scottish and English crowns are united (1603)
- Gunpowder Plot to blow up Parliament in Westminster is foiled (1605)

- Death of Tycho Brahe; the Danish aristocrat catalogued 777 stars and opened new paths for astronomy without accepting the Copernican system

- Santorio Santorio of Padua invents the *pulsilogium*, the first known device for taking the pulse

- Abraham Verhoeven, a printer in Antwerp, gets permission to print a weekly news periodical, the ancestor of the modern newspaper

- Galileo discovers the parabolic trajectory of projectiles; using a telescope he observes the Milky Way and phases of Venus, the mountains and craters of the Moon and the four largest satellites of Jupiter

- Johannes Kepler propounds his first two laws of planetary motion

◀ Newspaper published in London during the Great Plague of 1665

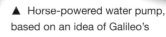

▲ Horse-powered water pump, based on an idea of Galileo's

1610

- Silver mining starts at Potosí Mountain in Bolivia (1611)
- The Romanov Dynasty ascends the Russian throne (1613)
- The Thirty Years' War breaks out in central Europe (1618)

- Santorio Santorio adds a gauge to the thermoscope, a device for measuring temperatures already described by Galileo, thereby creating the first thermometer

- The Spanish historian Antonio de Herrera y Tordesillas describes how the Aztecs used natural rubber, and is possibly the first to introduce the substance to Europe

- Nicolas Sauvage, a French carriage-maker, pioneers the idea of horse-drawn cabs for hire in Paris

1620

- *The Mayflower* arrives in New England (1620)
- New Amsterdam, the future New York, is founded (1626)
- Charles I attempts to rule Britain without calling a parliament (1629)

- Cornelis Drebbel builds the first submersible

- Dr William Harvey establishes that the heart functions like a pump and demonstrates that blood circulates around the body

◀ Mayahuel, the Aztec goddess of fertility, holds a sceptre decorated with fronds of the rubber tree

▼ Observations of cold-blooded animals such as fish helped William Harvey in his research to discover the secrets of blood circulation

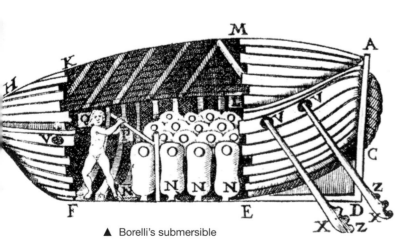

▲ Borelli's submersible

1630

- Work starts on the Taj Mahal in India (1631)
- The colony of Maryland is founded in America (1632)
- Japan cuts itself off from the outside world (1637)

- A French gunmaker named Marin Le Bourgeoys develops the flintlock musket, employing a firing mechanism that strikes sparks from flint

- Pierre Vernier demonstrates the calliper scale that still bears his name

- A 3 mile-an-hour speed limit is introduced for hackney cabs in London

- The umbrella makes its first appearance in the Europe

- The French mathematician and philosopher René Descartes publishes his *Discourse on Method*, introducing the principle of doubt as the measure of all things and announcing the coming of a 'universal science' based purely on reason; in other fields Descartes develops the Cartesian co-ordinate system and analytical geometry, and does important work in the field of optics, spelling out the law of refraction

1640

- Civil war breaks out in England (1642)
- The Manchu Qing Dynasty takes control of China (1644)
- The signing of the Treaty of Westphalia brings the Thirty Years' War to an end (1648)
- Charles I is executed (1649)

- Blaise Pascal invents his calculating machine, a distant ancestor of the modern computer

- Evangelista Torricelli invents the barometer

- Galileo dies under house arrest; he had made a major contribution to the development of science by insisting that precise measurement and formal logic, rather than abstract philosophical speculation, should be the foundation of physics, but his support for Copernicus's theory that the Earth revolves around the Sun – spelled out in his *Dialogue concerning the Two Chief World Systems* – led to him being accused of heresy by the Catholic Church in 1632

▲ Model of Pascal's calculating machine

► The perils of umbrellas

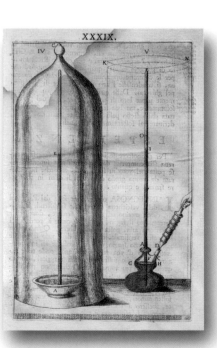

▲ Diagram illustrating atmospheric pressure, dating from the time of the invention of the barometer

1650

• The Dutch settle the Cape of Good Hope in South Africa (1652)
• Oliver Cromwell becomes Lord Protector of Britain (1653)
• Louis XIV is crowned king of France (1654)

• Grenades become a familiar weapon on European battlefields

• Otto von Guericke demonstrates the vacuum pump

• Jean-Jacques Renouard de Villayer installs wall-mounted post boxes in the main thoroughfares of Paris

• The Dutch scientist Christiaan Huygens invents the pendulum clock

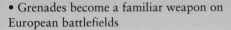

▶ Microscope attributed to Giuseppe Campani

1660

• Restoration of the British monarchy under Charles II (1660)
• China's Kangxi Emperor starts a 61-year reign (1662)
• Great Fire of London (1666)
• Aurangzeb bans the Hindu religion in India (1669)

• Melchisédec Thévenot invents the spirit level

• The Royal Society is founded in London in 1660; six years later the French Academy of Sciences has its first meeting in Paris

• Marcello Malpighi observes capillaries and founds the science of histology (the microscopic analysis of cell tissue)

• The first Western bank notes are issued in Sweden on the initiative of a Dutch banker, Johan Palmstruch

• Robert Hooke, a pioneer of microscopy, publishes his *Micrographia*, attracting a wide audience

• Paris's police commissioner, Gabriel Nicolas de La Reynie, makes a coordinated attempt to introduce street lighting in the city's main streets

• In the wake of the Great Fire, Nicholas Barbon sets up the first household insurance company in London

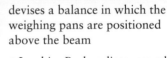

• Gilles Personne de Roberval devises a balance in which the weighing pans are positioned above the beam

• Joachim Becher discovers ethylene

▲ Venetian lamplighter

◀ Library of the Accademia dei Lincei, a scientific academy in Rome

1670

EVENTS

- The Hudson Bay Company is founded in North America to exploit the fur trade in the northern lands (1670)
- France establishes a colony at Pondicherry in India (1674), providing competition for British traders in the subcontinent
- War between Sweden and Denmark (1675–79)

INVENTIONS

- A Capuchin monk, Chérubin d'Orléans, produces a prototype pair of binoculars

- Isaac Newton invents the reflecting telescope

- Christiaan Huygens patents a watch with a spiral spring mechanism

- Using microscopes of his own devising, Anton van Leeuwenhoek discovers red corpuscles in the blood and studies spermatozoa; he also studies yeasts in beer and notes the presence of micro-organisms (protozoa and bacteria) in stagnant water

- George Ravenscroft takes out a patent on lead glass

- Edmund Halley draws up the first sky map of the Southern Hemisphere

1680

EVENTS

- An Ottoman army unsuccessfully besieges Vienna (1683)
- China annexes Taiwan (1683)
- Louis XIV revokes the Edict of Nantes (1685)
- William and Mary take the British throne as joint monarchs (1689)

INVENTIONS

- John Ray publishes the first part of his *Historia Plantarum*, adopting a scientific approach to the classification of plants

- Ice cream is served to King James II, its first recorded appearance in Europe; by 1686 the Procope café in Paris is specialising in the delicacy

- Edmund Halley publishes the first meteorological map of the world, indicating the prevailing winds over the oceans

- Denis Papin invents the pressure cooker

- Isaac Newton publishes the *Philosophiae Naturalis Principia Mathematica* ('Mathematical Principles of Natural Philosophy'), expounding the law of universal gravitation and founding modern astrophysics; a mathematician and astronomer as well as a physicist, Newton goes on to develop infinitesimal calculus (a discovery made independently by the philosopher Gottfried Leibniz), propounds the generalised binomial theorem and demonstrates that a prism can separate white light into a spectrum of colours

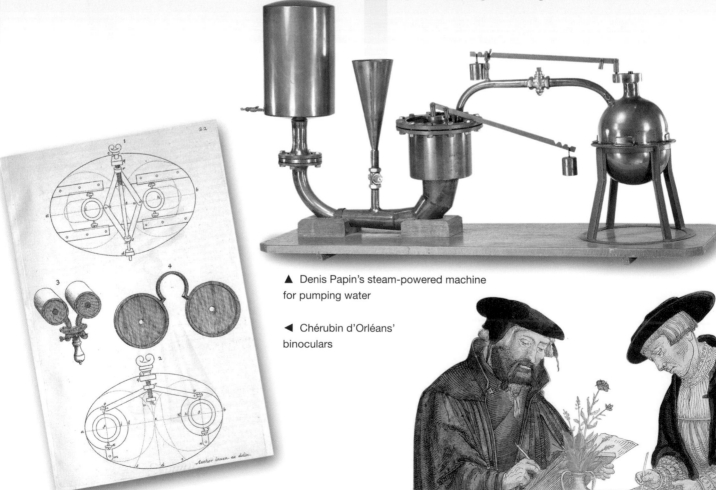

▲ Denis Papin's steam-powered machine for pumping water

◄ Chérubin d'Orléans' binoculars

▲ Two artists prepare illustrations for an early botanical work

1690

- Witch trials begin in Salem, Massachusetts (1692)
- Potala Palace completed in Lhasa, Tibet (1694)
- Bank of England founded (1694)

- Edmund Halley constructs a diving bell for underwater research

- Dom Pérignon, cellarer of the Benedictine abbey of Hautvilliers in northeastern France, helps to develop champagne

- Death of Christiaan Huygens, a mathematician and physicist responsible for important astronomical discoveries including the rings of Saturn and Saturn's largest satellite Titan, the rotation of Mars, the Orion nebula and the transit of Mercury; other contributions include devising the principle of the pendulum clock, coining the term 'centrifugal force', studying logarithmic (exponential) curves, and developing the wave theory of light

- Thomas Savery patents a water pump, marking the first practical application of steam power

1700

- Foundation of St Petersburg (1703)
- The Act of Union formally joins England and Scotland within Great Britain (1707)
- The death of Aurangzeb signals the onset of the lengthy decline of India's Mughal Empire (1707)

- Johann Christoph Denner introduces improvements to the medieval chalumeau, creating the modern clarinet

- Jethro Tull invents the seed drill

- Edmund Halley publishes his *Synopsis of the Astronomy of Comets*, in which he applies Newton's laws of motion to show that three comets sighted separately in 1531, 1607 and 1682 were actually one and the same, following a trajectory in accordance with the theories stated in the *Principia Mathematica*

- Bartolomeo Cristofori invents the piano

- An Italian perfumer based in Cologne develops a new fragrance and calls it *eau de Cologne*

1710

- The Peace of Utrecht brings the War of the Spanish Succession to an end (1713)
- George I becomes King of England, founding the Hanoverian Dynasty (1714)
- Death of Louis XIV of France (1715)

- John Shore invents the tuning fork

- Jacques Bernoulli lays the foundations of the theory of probabilities and of the science of statistics, building on earlier work by Pascal and the mathematician Pierre de Fermat

- Martin Triewald has the idea of circulating hot water in covered channels to heat a greenhouse, marking the start of modern central heating

- Jacob Cristoph le Blon develops 4-colour printing to produce coloured reproductions of paintings

- The alchemist J F Böttger manages to replicate some of the secrets of Chinese porcelain by using kaolin to make pottery (1709); Augustus the Strong of Saxony establishes a manufactory in Meissen the following year

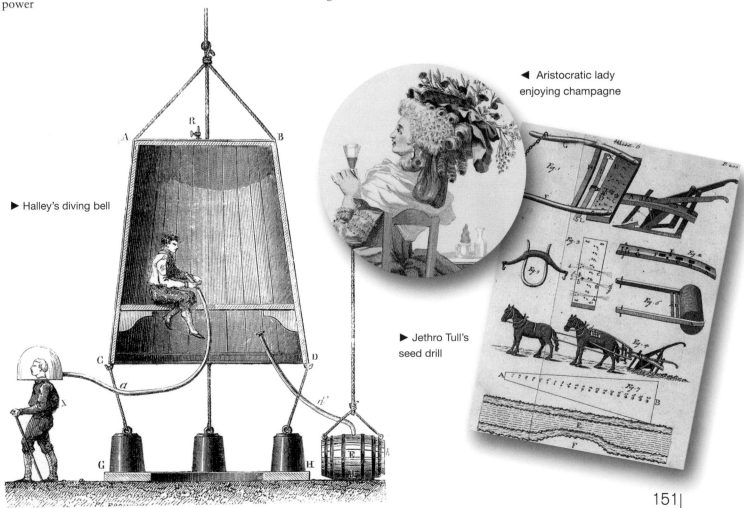

► Halley's diving bell

◄ Aristocratic lady enjoying champagne

► Jethro Tull's seed drill

1720

1730

1740

EVENTS

- South Sea Bubble causes financial crash in England (1720)
- The Peace of Nystadt ends the Northern War between Russia and Sweden (1721)
- Vitus Bering explores the Bering Strait between America and Asia (1728)

- The colony of Georgia is founded – the last of the 13 colonies that will one day inaugurate the USA (1732)
- Nadir Shah becomes ruler of Persia (1736)
- The last Medici Grand Duke of Tuscany dies (1737)

- Frederick II, the Great, becomes king of Prussia (1740)
- In Arabia, the Saudi sheikhs adopt Wahhabism (1740)
- The War of the Austrian Succession divides Europe (1740–1748)
- Jacobite Rebellion ends at the Battle of Culloden (1746)

INVENTIONS

- Catherine I founds the Imperial Academy of Sciences in St Petersburg
- Death of Sir Isaac Newton (1727)

- John Hadley develops the principle of the marine sextant
- Carl von Linné (Linnaeus) publishes his *Systema Naturae,* introducing the system of classifying animals and plants by a two-part Latin name listing first the genus, then the species
- Jacques de Vaucanson makes his pioneering automaton *The Flute Player*

- William Brownrigg makes the first non-natural sparkling water
- Ewald Jürgen von Kleist creates the first electrical capacitor, known as a Leyden jar
- Andreas Sigismund Marggraf, a German apothecary, produces sugar from sugar beet

◀ Linnaeus's classification of mushrooms

▶ *The Draughtsman,* an automaton made by the Jacquet-Droz brothers

1750

- Death of the composer Johann Sebastian Bach (1750)
- The Lisbon earthquake kills 30,000 people (1755)
- The Seven Years' War breaks out affecting Europe, North America and India (1756)

- The first volume of the *Encyclopédie,* a key work of the Enlightenment, is published in France

- Benjamin Franklin invents the lightning conductor

- The British Museum opens in London, becoming the first free national collection

1760

- The Treaty of Paris ends the Seven Years' War, confirming Britain's position as Europe's dominant colonial power (1763)
- James Cook sets sail on his first Pacific voyage (1768)
- Catherine the Great launches the First Russo-Turkish War (1768)

- Jean-Joseph Merlin, a Belgian inventor, patents roller skates

- Giovanni-Battista Morgagni develops the discipline of anatomical pathology

- John Montagu, the 4th Earl of Sandwich, popularises the sandwich, which comes to bear his name

- John Spilsbury, a London mapmaker, produces the first jigsaw puzzles

- The ironmaking plant at Coalbrookdale in Shropshire makes the first cast-iron rails for wagonways, the predecessors of railways

- A former dragoon, Philip Astley, opens a trick-riding arena in London that becomes the prototype of the modern circus

◀ Priestley's method of carbonating water

▶ Sugar beet

▼ Johann Sebastian Bach playing for Frederick the Great of Prussia

Index

Page numbers in *italics* refer to captions.

A

abacuses 48
Abbe, Ernst 71
Academia Naturae Curiosi 125-6
academies 125-6
Accademia dei Lincei 38, 68, 127, *149*
Accademia del Cimento 127
Achard, Franz Karl 138
Achenwall, Gottfried 130
actuarial calculations 77, 117
Addison, Joseph 21
agriculture
    agricultural revolution 115
    crop rotation 115
    land enclosures 115
    seed drills 114-15, *114*, 151
    steam-powered threshers 112, *112*
alchemy 97, *105*
Alembert, Jean le Rond d' 128
Alexander the Great 33
algebra 57
All Souls College, Oxford *4*
almanacs 20-1
Alpago, Andrea 40
American War of Independence 34, *34*, *134*
Amsterdam *17*, 21, 140-3, *141*
analytical engine 53
analytical geometry 57
anatomy *14*, 102-4, 153
    anatomical theatres 42, *103*, *124*
    dissections *14*, 41, 103, *103*, 104
androids 136, *136*
animism 107
Antikythera mechanism 50, *50*
Apollonius of Perga 118
Arab and Islamic scholars 28, 30, 40-1, 99
Archimedes 50
Aristotle 36, 40, 54, 93, 94
    cosmology 27, 30, 37
arithmometers 53, *53*
armillary spheres *7*, *31*
arquebuses 44, 46
Astley, Philip 153
astronomical globe *30*
astronomy 27-31, 116-18
    Arabic 28
    Aristotelian system 27, 30, 37
    armillary spheres *7*, *31*
    Cartesian 54, *55*, 57

comets 83, 116-18, *117*, *118*, 151
Copernican system *7*, 27, *27*, 28-9, *29*, 30-1, 38, 39, 118
    ephemerides 135
    Galilean system 37-8
    heliocentrism 27, *27*, 28, *29*, 30, *31*, 54
    Moon 37, 83, 94, 95, *95*, 146
    planets 37, 80, *80*, 82, 94, *94*, 117, 146
    Ptolemaic system 28, *28*, 30
    stars 80, 85, 88
    see also telescopes
atmospheric pressure 9, 49, 58-9, 61
Austria 122
automatons 136, *136*, 152
Aztecs 24, *24*

B

Babbage, Charles 53, *53*
Babcock, Alpheus 123
Baccarat 89
Bach, Johann Christian 122
Bach, Johann Sebastian 121, 153
Bacon, Roger 21, 30
bacteriology 71
balance scales 79, *79*, 149
Bank of England 64, 66, 151
Bank of France 65
Bank of Scotland 64
bankers' drafts 63-4, 142
banknotes *10*, 62-6, *62*, *63*, *64*, 149
Barbon, Nicholas 76, 149
Barlet, Annibal *105*
barometers 9, 49, 58-9, *58*, *59*, 148
    dial barometer 68
Barrow, Isaac 82
Barthez, Paul Joseph 107
Barton, Sir Henry 78
bathyschaphes 35
bayonets 46, *46*
Becher, Joachim 149
Bechet, Sidney *119*
Beethoven, Ludwig van 122
Belgium 61, 105, 130
benevolent societies 77
Béranger, Joseph 79

Bérié, Jean 32
Berlin Academy of Sciences 127, 138
Berlioz, Hector *122*
Bernary, Ernst *138*
Bernoulli, Daniel 127
Bernoulli, Jacques 130, 151
Bernoulli, Nicholas 127
bestiaries 93
bills of exchange 63-4
binary system 100
binocular microscopes *71*
binoculars 88, 150
    see also telescopes
binomial coefficients 49
biomechanics *14*
Blake, William *99*
blood circulation 7, 8, *8*, 40-3, 104, *107*, 147
blood transfusions 42
blood-letting 8, 43, 105
Bock, Hieronymus 90
Boehm, Theobald 119
Boerhaave, Hermann 106, *124*
Bonnemain, Jean Simon 132
Bopp, Thomas *117*
Bordeu, Théophile de 107
Borelli, Giovanni Alfonso *14*, 32, *32*, 106, *106*
Bossuet, Jacques-Benigne 56
botanical gardens 92, 141, 146
botany *13*, 90-3
    herbals 91, *91*
    Linnaean nomenclature 92-3
    plant classification *13*, 90-3, 150
Böttger, J F 151
Boulton, Matthew 111
Bourne, William 32
Boyle, Robert 61, 68, 109, 113
Brahe, Tycho 28, 85, 146
Brandenburg Society of Sciences 126
Britain *4*, 6, 21, 23, 26, 33, 34-5, 41-3, 47, 51, 52, 53, 61, 64, 68, 71, 76, 77, 78, 79, 82, 84-5, 86, 89, 90, 92, 95-100, 109-12, 114-18, 122, 126, 127, 128, 129, 134-5, 137
British Museum 128, 153
broadsheets 20
Broadwood, John 122
Brown, George 52
Brown, Joseph 47
Brownrigg, William 137, 152
Brueghel, Jan, the Younger *142*
Brun, Charles 34
Brunfels, Otto 91

Bruno, Giordano 5, 27, 30, 37, 146
Buffon, Comte de 93
Buridan, Jean 30
Burke, Edmund 23
Bushnell, David 32

C

cabinets of curiosities 127-8
Caesalpinus, Andreas 41
Cage, John 123
calculating machines 9, 13, 48-53, *48*, 148
    analytical engine 53
    Antikythera mechanism 50, *50*
    arithmometer 53, *53*
    calculating clock 49
    difference engine 53, *53*
    pascaline *48*, 49-52
    slide rule 51, *51*
    stepped reckoner 52, *52*
callipers 47, *47*, 148
Calvin, John 124
Cambridge University 125, *125*
Camerarius, Rudolf 91
Campani, Giuseppe 68
Campbell, John 135
capacitors 137, 152
carbonated drinks 137, *137*
Carnot, Sadi 113
Cartesian coordinates 57
Cartesians 55, 56
    see also Descartes, René
Cartwright, Edmund 110
Cassegrain, Laurent 84-5
Cassini, Giovanni 80
Cassini, Jean-Dominique 116
Cassini-Huygens space probe *81*
cast iron 113, 153
Catherine I of Russia 126, 152
Catherine the Great 127, 153
Catholic Church 4, 9, 27, 31, 38, 39, 124
Cawley, John 110
cell division 70
cellular biology 71
censorship
    academic self-censorship 127
    of the press 23
    religious 39
central heating 132, 151
centrifugal force 80
Cesalpino, Andrea 90-1
Cesi, Federico 127
chalumeau 119
champagne 101, 151
Charles X of France *30*

Cheere, John 97
Chelsea Physic Garden 92
cheques 66
Chérubin d'Orléans 88, 150
China 20, 47, 60, 63, 78, 89, *113*
choc-ice bars 89
Chopin, Frédéric 122
Christina, Queen of Sweden 56, *56*, 62
chromatic aberration 68, 70, 71, 82, 84
chromolithography 133
chromosomes 70
circulatory system 7, 8, *8*, 40-3, 104, *107*, 147
circuses 153
cladistics 93
Clairaut, Alexis 118
clarinets *16*, 119, *119*, 151
clavichords 120, 121
clocks
    pendulum clocks *12*, *39*, 80, *81*, 149
    watches 150
Coalbrookdale *110*, 153
coinage 62, *62*, 63
Colbert, Jean-Baptiste 81, 126, *126*, 127, 142
colimpha 33, *33*
Colmar, Charles-Xavier Thomas de 53
Colombus, Renaldus 41, 42, 102
Coltelli, Procopio dei 89
Columbus, Christopher 24, 30, 138
combustion 68
comets 83, 116-18, *117*, *118*, 151
Condorcet, Nicolas de 128
Continental System 139
Copernicus, Nicolas 7, 27, *27*, 28-9, *29*, 30-1, 38, 39, 118
Coryat, Thomas 47
Coster, Salomon *81*
counterfeiting 63
Crimean War *46*, 60
Cristofori, Bartolomeo *16*, 120-1, 151
crop rotation 115
crystal 89
crystallography 71
Cugnot, Nicolas-Joseph 112
Culloden 60

D

Danhauser, Josef *122*
Darwin, Charles 93
Daudeleau rifles 46
De revolutionibus orbium coelestium 7, 28, 30, 31

Debussy, Claude 123
decompression chambers 33
Delessert, Benjamin 139
Demisiano, Giovanni 68
Denner, Johann Christoph *16*, 119, 151
Denner, Johann David 119
Denys, Jean-Baptiste 42
Descartes, René 5, 10-11, *10*, 43, 52, 54-7, *54*, *56*, 80, 84, 99, 105, 126, 142, 148
determinism 130
Deyeux, Nicolas 139
dial barometers 68
Dialogue Concerning the Two Chief World Systems 38, 39
Diderot, Denis 128
diesel engines 35
difference engine 53, *53*
digestion 105
Digges, Sir Thomas 30
Dioscorides 91
Discourse on Method 54, 148
Discourses and Mathematical Demonstrations Relating to Two New Sciences 39
dissections *14*, 41, 103, *103*, 104
diving bells 33, 117, 151
Dockwra, William 61
Dolland, John 86
Drebbel, Cornelius 6, 32, *32*, 67-8, 147
Du Deffand, Madame 128
Duillier, Fatio de 99
dulcimers 121
Dumesnil, Louis-Michel *56*
Dürer, Albrecht 91
Dutch East India Company 141, *141*, 146
Dutch West India Company 141

E

E-tuners 129
eau de Cologne 132, *132*, 151
Edict of Nantes 109, 150
education 124-8, 143
    academies 125-6
    universities *4*, *16*, 37, 42, 102, 124, 125, 126
Egypt, ancient 83
Einstein, Albert 38, 96, 99, 130
elasticity, law of 68
electricity 113, 137
electron microscopes 71, *71*
electronic money 66
Elizabeth of Bohemia *56*
employment agencies 22

Encyclopédie *42*, 128, 136, 153
Enfield rifles 46
Engelmann, Godefroy 133
Enlightenment *4*, 124-8
ephemerides 135
Equilbey, Laurence 129
Érard, Sébastian 123
erasers 25
Eratosthenes 7
ergonomics 50
ethylene 149
Euclid 49
euro 66
Eustachi, Bartolomeo 104
evolutionary theory 93

F

Faber, John 68
Fabricius, Hieronymus 41, 42, 102, *103*, 104
Fallopius, Gabriel 42, 102
Faques, Richard 20
al-Farghani 30
Farina, Johann Maria 132
Feminis, Giovanni Paolo 132
Fénelon, François 56
Fermat, Pierre de 49, 130
fermentation 71, 105
fire insurance 76-7
fire pumps 143
firearms 44-6
    arquebus 44, 46
    bayonets 46, *46*
    flintlock musket 7, *44*, 45, *45*, 46, 148
    matchlock musket 44, *44*, 45
    percussion caps 46
    rifles 46, *46*
    wheel-lock musket 44, 45, *45*
Flamsteed, John 116
Fleury, Cardinal de *133*
flintlock muskets 7, *44*, 45, *45*, 46, 148
Flodden, Battle of 20
flush toilets 146
Fontana, Francesco 68
Fontenelle, Bernard le Bovier de *94*, *128*
food and drink
    champagne 101, 151
    ice cream 89, *89*, 150
    sorbets 89
    sparkling water 137, *137*, 152
    sugar manufacture *17*, 138-9, *138*, *139*, 152
force of attraction 96
Ford, Brian J 75

Forsyth, Alexander 46
Foucault, Léon 86-7
four-colour printing 133, *133*, 151
fourth estate 23
France 22, 23, 26, 32, 34, 44, 47, 48-52, 54-7, 61, 65, 78, 79, 84, 86-7, 88, 89, 92, 93, 101, 107, 108, 112, 113, 125, 126, 130, 136, 138-9
Franklin, Benjamin 128, 135, 153
Frederick I of Prussia 75
Frederick III of Prussia 126
French Academy of Sciences 79, 81, 85, 126, *126*, 127, *127*, 138, 149
French Revolution 22, 65, *65*
Fresneau, Francois 24-5
Fuchs, Leonard 90, 91, *91*
Fulton, Robert 34-5

G

Galen 40, 41, 43, 102
Galileo Galilei 5, *8*, 30, 36-9, *36*, 54, 58, 67, 68, 80, 82, 83, 94, 96, 98, 105-6, 127, 146, 147, 148
Garrett, Revd George 34-5
gas lighting 78
Gassendi, Pierre 126
Gauss, Carl Friedrich 130
Gaussian function 130, *130*
gelatin 113
genetics 71
Germany 20, 47, 49, 52, 61, 71, 83, 91, 93, 100, 107, 119, 121, 122, 123, 126, 130, 133, 137, 138
Gesner, Conrad 90, 93, *93*
Gibbs, James *4*
Gingerich, Owen 31
Giustini, Lodovico 121
glass
    lead glass 89, 150
    plate glass 89, *89*
Godfrey, Thomas 135
gold standard 66
Goodyear, Charles 25
Graaf, Régnier de 73, 105
gravitation, theory of 12, *14*, 30, 39, 68, 81, 94-6, 98, 118
Great Fire of London 76, *76*
Greece, ancient 36, 40, 41, 43, 50, 54, 77, 91, 93, 102, 118
Gregorian telescope 68
Gregory, James 82, 84, 85
grenades 60, *60*, 149
Guericke, Otto von 61, 108, 149

gunpowder 60
Gunter, Edmund 51
Gutenberg, Johannes 21

H

hackney carriages 26, 148
Hadley, John 17, 134, 135, 152
Hale, Alan 117
Hale-Bopp comet 117
Hall, Chester Moore 71
Halley, Edmund 15, 33, 77, 116-18, 116, 150, 151
Halley's comet 87, 117, 118
Hals, Frans 54
Hammurabi of Babylon 76-7
Hanseatic League 141
Hansom, Joseph 26
Hanway, Jonas 47
harpsichords 120
Harrington, Sir John 146
Harrison, John 135, 135
Hartsoeker, Nicolas 70, 75
Harvey, William 7-8, 7, 8, 40, 40, 41-3, 43, 74, 104, 105, 147
Haupt, Hans 47
Hawksmoor, Nicholas 4
Haydn, Joseph 122
Hegel, Georg Wilhelm Friedrich 56
Heisenberg, Werner 130
helioscopes 68
Helmont, Jan Baptist van 105
Henri IV of France 44, 45, 146
herbals 91, 91
Hérissant, Jean-Thomas 25
Herrera y Tordesillas, Antonio de 24, 147
Herschel, William 86
Hevelius, Johannes 13, 83, 83
Hippocrates 102
histology 149
Historia Plantarum 90, 90
Hoffmann, Friedrich 107
Holland 20, 69, 70, 72, 80, 83, 105, 106, 140-3
Holland, John 35
Holland VI 35
homunculi 70, 105
Hooke, Robert 11, 67, 67, 68, 68, 69, 69, 70, 72, 73, 75, 99, 149
horse-drawn cabs 6, 26, 26, 147
Hortus Botanicus 141
Hubble Telescope 87, 87, 96
humanism 9
humours, theory of 102, 102, 105
Hunter, John 107

Huygens, Christiaan 12, 74, 77, 80-1, 80, 82, 108, 127, 149, 150, 151
Huygens, Constantijn 80
hypocausts 132

I

Ibn al-Hasan (Alhacen) 99
Ibn al-Nafis 40-1
ice cream 89, 89, 150
Imperial Academy of Sciences 126-7, 152
Industrial Revolution 15, 66, 77, 108, 110, 128
inertia, principle of 30, 37, 81
infectious diseases 107
infinitesimal calculus 100, 150
Inquisition 27, 38
insurance 76-7, 149
internal combustion engines 35, 113
Italy 20, 32, 33, 36-9, 58-9, 68, 70, 77, 83, 88, 90-1, 92, 96, 102, 103, 105-6, 120-1, 127, 132

J

Jacquet-Droz brothers 136, 136
Jansen, Zacharias 69
Jansenism 52
Jardin des Plantes 92
Jewish communities 143
jigsaw puzzles 153
John Paul II, Pope 39
journalism *see* newspapers
Jupiter 37, 94, 116, 117, 146

K

Kassatkine, N A 112
Kepler, Johannes 30-1, 49, 94, 94, 146
kinematics 39
Kircher, Athanasius 28
Kleist, Ewald Jürgen von 137, 152
Koestler, Arthur 31

L

La Condamine, Charles-Marie de 24
La Gazette 22
La Reynie, Gabriel Nicolas de 78, 149

Lake, Simon 35
Lamarck, Jean-Baptiste 93, 93
lampoons 20
land enclosures 115
Lang, Fritz 136
Lang Lang 123
Laplace, Pierre-Simon 130
latitude and longitude 116, 134-5
Law, John 65
Le Blon, Jacob Christoph 133, 133, 151
Le Bourgeoys, Marin 9, 44, 45, 148
Le Boursier de Coudray, Angélique-Marguerite 106
lead glass 89, 150
Leaning Tower of Pisa 36
Lebel rifles 46
Lee, Ezra 34
Leeuwenhoek, Anton van 11, 12, 69, 70, 72-5, 72, 75, 105, 150
Leibniz, Gottfried 48, 52, 52, 56, 74, 126
Leonardo da Vinci 33, 91
Leoni, Ottavio 36
Lespinasse, Julie de 128
Leyden jars 137, 152
Leyden University 16, 124
Liebniz, Gottfried Wilhelm 100, 100
light
     light waves 81, 99
     white light 84, 99, 150
lightning conductors 153
Linnaeus, Carl 92-3, 92, 152
Lippershey, Hans 69, 83
Liszt, Franz 122, 122, 123
literacy 20
Lloyd, Edward 77
Lloyds of London 76, 77
logarithms 51, 80
looms 110
Louis XIV of France 78, 81, 101, 126, 151
Lovelace, Lady Ada 53
Lower, Richard 42
Luther, Martin 124

M

Machiavelli, Niccolò 125
MacIntosh, Charles 25
Macquer, Pierre-Joseph 25
Maffei, Scipione 121
magnetic declinations 15, 117
Maine, Duchess of 128, 128
malaria 104
Malaysia 25
Malebranche, Nicolas 56
Malpighi, Marcello 42, 43, 70, 75, 103, 105, 149

maps
     Cartesian coordinates 57
     cartographic projection techniques 116
     meteorological maps 117, 150
     sky maps 116, 116, 150
Marchi, Francesco de 33
Marggraf, Andreas Sigismund 17, 138, 152
marine chronometers 135, 135
Mars 80
Martin, Benjamin 70
Martire d'Anghiera, Pietro 24
Mascagni, Paolo 104
matchlock muskets 44, 44, 45
mathematics 5, 36, 54, 55, 57
     algebra 57
     analytical geometry 57, 148
     binary system 100
     infinitesimal calculus 100, 150
     logarithms 80
     Pascal's triangle 49
     statistics 130-1
     superscripts 57
     see also calculating machines
Maupertuis, Pierre Louis 127
Mayer, Tobias 118
mechanics 36, 106
Medhurst, John 79
Medici, Leopoldo de 127
medicine 102-7
     blood transfusions 42
     circulatory system 7, 8, 8, 40-3, 104, 107, 147
     humours, theory of 102, 102, 105
     infectious diseases 107
     midwifery 106
     theories of the human body 105-7
     see also anatomy; microscopy
Menabrae, Frederico Luigi 53
Merlin, Jean-Joseph 153
Merret, Christopher 101
Mersenne, Marin 32, 56, 126
Mesmer, Franz Anton 128
Mesopotamia (Iraq) 47
meteorological maps 117, 150
Metropolis 136
mezzotint process 133
microbiology 72, 75, 107
     see also microscopy
Micrographia 11, 67, 67, 68, 70, 73
microscopy 11, 12, 67-75, 68, 69, 104-5, 149, 150
     binocular microscope 71
     chromatic aberration 68, 70, 71
     compound microscope 69

electron microscope 71, *71*
ultraviolet microscope 71
midwifery 106
Milky Way 146
Mills bombs 60
Molière 22, 105, 128
money
  bankers' drafts 63-4, 142
  banknotes 10, 62-6, *62, 63, 64*, 149
  bills of exchange 63-4
  cheques 66
  coinage 62, *62*, 63
  counterfeiting 63
  electronic money 66
  euro 66
  speculation 142
Montesquieu, Charles de Secondat, baron de *128*
Monturiol, Narcisse 34
Moon 37, 83, 94, 95, *95*, 146
Morgagni, Giovanni Battista 103, 153
Mozart, Wolfgang Amadeus 122, *122*
Müller, Ivan 119
Müller, Johannes 30
musical instruments
  chalumeau 119
  clarinet *16*, 119, *119*, 151
  clavichord 120, 121
  dulcimer 121
  harpsichord 120
  piano *16*, 120-3, *121, 122, 123*, 151
  spinet *120*
musical pitch 129
musical theory 126
Musschenbroek, Pieter van 137

N

Napier, John 51, *51*
Napoleon Bonaparte 34, 65, 138-9, *139*
natural selection 93
Nautilus 34
navigation 17, 134-5
  see also maps
Necker, Madame 128
Nehou, Louis Lucas de 89
Nelson, Christian 89
Newcomen, Thomas 110
news-sheets 20, *21*
newspapers 6, 20-3, *142*, 143, 146
Newton, Isaac 5, 12-13, *12, 13, 14*, 30, 31, *55*, 57, 68, 81, 82, 84, *84*, 85, 86, 95-6, 97-100, *97*, 117, 118, 126, 150, 152

Nicholas of Cusa 30
Nieuwe Tijdinghe ('New Tidings') 6, 20
night-vision binoculars 88

O

observatories 87
octants *17, 134*, 135
Oldenburg, Henry 73, 74
opera glasses 88
optics *12*
  binoculars 88
  refraction, law of 57, 148
  see also light; telescopes
Oresme, Nicole 30
Oughtred, William 51
Oxford University 4, 125, *125*
Oxford University Botanic Garden 92

P

Paganini, Niccolò 122
Palmstruch, Johan 62, 63, 64
pamphlets 20
Pape, Henri 122
paperboys 23
Papin, Denis 14, *15*, 108-9, *108*, 113, 150
Parra, Felix 39
Parsonstown Leviathan 86
Pascal, Blaise 9, 13, 48-9, *49*, 50-2, 56, 77, 126, 130, 148
Pascal, Etienne 48-9
Pascal's triangle 49
pasclines *48*, 49-52
Pasteur, Louis 71, 107
Peal, Samuel 25
pendulums *12*, 36-7, 39, 80, *81*, 149
penny post 61
Pepys, Samuel 97
percussion caps 46
perfume 132
Pérignon, Pierre 101, 151
periscopes 88
Peter the Great of Russia 75
philosophy 10, 54, 55-6, 57
physiology 57, 105, 106-7
pianos *16*, 120-3, *121, 122, 123*, 151
Piccard, Auguste 35
Pico della Mirandola, Giovanni 125
pie charts and bars *131*
pillar boxes 61
Pirelli *25*
pitch pipes 129
plague 21, 100, 107
Planck, Max 99

planets 37, 80, *80*, 82, 94, *94*, 117, 146
plant classification *13*, 90-3, 150
plate glass 89, *89*
Plato *28*
Pliny the Elder 93
Poinsot, Louis 79
Poland *28*, 30
Politian 125
Polo, Marco 63
Pope, Alexander 21
porcelain 151
Porro, Ignazio 88
post boxes 61, *61*, 149
postal services 20, 61
precision measurements 47
press censorship 23
pressure cookers 113, 150
Price, Richard 77
Priestley, Joseph 25, *70*, 137, *137*
Principia Mathematica 95, *96*, 98, *100*, 118, 150
Principles of Philosophy 10, 54, 57
printing presses *21*
  see also newspapers
probabilities, theory of 49, 130
Prokofiev, Sergei 123
Prussia 23, 77
  see also Germany
Ptolemy *28, 28*
publishing industry 143
  see also newspapers
pulse measurement 37, 146
pumps
  fire pumps 143
  steam pumps *15*, 109, *109*, 110, *110*, 151
  vacuum pump 61, *61*, 149
  water pumps 37, *37*, 109, 110, 111, 151

Q

Quételet, Adolphe 130-1
quinine 104

R

Rachmaninov, Sergei 123
Radcliffe Camera *4*
railways 112-13, *113*
Ramsden, Jesse *85*
rationalism, Cartesian 52, 54, 55, 56, 57, *57*
Ravel, Maurice 123
Ravenscroft, George 89, 150
Ray, John 90, *90*, 91, *91*, 150
red blood cells 74

reflecting telescopes 68, 82-7, *82, 85*, 150
Reformation 9, 124, 125
refracting telescopes *13*, 82-3, *83*, 86, 87
refraction, law of *57*
relativity 38, 96
Rembrandt van Rijn 14, *104, 140*, 143, *143*
Renaissance 4
Renaudot, Théophraste 22
Rheticus 30
Richelieu, Cardinal 22, 48
rifles 46, *46*
Rigaud, Hyacinthe *133*
Riolan, Jean 43
Roberval balance 79, *79*, 149
Roberval, Gilles Personne de *12*, 79, 149
Roebuck, John 111
roller skates 153
Rossini, Gioachino *122*
Rousseau, Jean-Jacques 23
Royal Prussian Academy of Sciences 126
Royal Society 67, 68, 73, 75, 81, 82, 85, 100, 111, 113, 127, 134, 135, 149
rubber 24-5, *24, 25*, 147
Russia 119, 126-7

S

salons 128
Sand, George *122*
sandwiches 153
Santorio Santorio 146, 147
Saturn 37, 80, *80, 81*, 82, 94, 116, 117
Saurin, Joseph *128*
Sauvage, Nicolas 6, 26, 147
Savery, Thomas *15*, 109-10, 151
scanning electron microscopy *71*
Schickard, Wilhelm 49
Schleiden, Matthias Jakob 93
scholasticism *8*, 124
Schubert, Franz 122
Schumann, Robert 122
Schwann, Theodor 93
Scientific Revolution 5, 7, 52, 99, 125
Scultetus, Johannes *42*
seed drills 114-15, *114*, 151
Serre, Michel *107*
Servetus, Michael 41
sextants 134-5, *135*, 152
Shore, John 129, 151
Shovell, Sir Cloudesley 134
Siegen, Ludwig von *133*
Silbermann, Gottfried 121, 123

Silvaticus, Matthaeus 92
sky maps 116, *116*, 150
slave trade 141
slide rules 51, *51*
Sloane, Sir Hans 128
Smeaton, John 33
Snow, Peter *131*
social security 77
sociology 131
sorbets 89
South America 24-5
Spain 34
Spallanzani, Lazzaro *68*
sparkling water 137, *137*, 152
Spectator 21
sperm 74, 105
Spilsbury, John 153
spinets *120*
Spinoza, Baruch 56
spirit levels 78, *78*, 149
spring scales 79
Stahl, George 107
stars 80, 85, 88
statistics 130-1, 151
   Gaussian function 130, *130*
   probabilities 49, 130
   theory of errors 130
steam engines 14-15, *15*, 108-13, *108*, *115*
steam pumps *15*, 109, *109*, 110, *110*, 151
steam ships and boats 35, 109, *109*, 113
steam trains 112, *113*
steam-powered threshers 112, *112*
Steele, Richard 21
Stein, Johann Andreas 122
Steinway and Sons 123
stepped reckoners 52, *52*
stock exchanges 142, *142*
Storck, Abraham *141*
street lighting 78, *78*, 143, 149
Streicher, Nanette 122
Stukeley, William 95
submersibles 6, 32-5, 147
   bathyschaphes 35
   colimpha 33, *33*
   diving bells 33, 117, 151
   submarines 32, *32*, 34-5, *35*
sugar beet *17*, 138-9, 152
sugar manufacture *17*, 138-9, *138*, *139*, 152
sundials *98*
sunspots *83*
supernovas 37, 85
superscripts 57

Swammerdam, Jan 70
Sweden 62-3
swingometers *131*
Switzerland 93, 130, 136
Sylvius, Franciscus 105

T
taxis 26
telescopes *8*, *13*, 30, 37, *38*
   angular resolution 87
   binocular telescope 88, *88*
   chromatic aberration 82, 84
   Gregorian telescope 68
   Hubble Telescope 87, *87*, *96*
   observatories 87
   Parsonstown Leviathan *86*
   reflecting telescope 68, 82-7, *82*, *85*, 150
   refracting telescope *13*, 82-3, *83*, 86, 87
   Schmidt-Cassegrain telescope 85, *85*
   see also binoculars
textile industry 110, *112*
thermodynamics 113
thermometers 68, 147
Thévenot, Melchisédech 78, 149
Thomas Aquinas, St 54
The Times 23
Titan 80, *81*
Torricelli, Evangelista 49, 58, 59, 108, 127, 148
Tournefort, Joseph Pitton de *13*, *91*, 92
transmutation, concept of 93
Triewald, Martin 132, 151
Trollope, Anthony 61
tulipmania 142, *142*
Tull, Jethro 114-15, 151
tuning forks 129, *129*, 151
Turtle 32, 34, *34*
al-Tusi, Nasir ad-Din 28, 30
tyres 25

U
ultraviolet microscopes 71
umbrellas 47, *47*, 148
uncertainty principle 130
United States 21, 23, 34, 35, 77, 123, 135
universities *4*, *16*, 37, 42, 102, 124, 125
Urban VIII, Pope 38

V
vacuum pumps 61, *61*, 149
Vaillant, Bernard 80
Vaucanson, Jacques de 136, 152
Venus 37, 146
Verhoeven, Abraham 20, 146
Vermeer, Jan *11*, 75, *75*
Verne, Jules 34
vernier calliper 47, *47*, 148
Vernier, Pierre 47, 148
Versailles, Palace of *89*
Vesalius, Andreas 41, 42, 102
Villayer, Jean-Jacques Renouard de 61, 149
Villiez, Jean-François *64*
vitalism 107
vivisection 41, *56*
Voltaire *128*
vulcanisation 25

W
Walter, Anton 122
watches 150
water pumps 37, *37*, 109, 110, 111, 151
watermarks 63
waterproof fabrics 25
Watt, James 110, 111
Watteau, Antoine *101*
weapons
   grenades 60, *60*
   see also firearms
Weiditz, Hans 91
weighing scales *12*
   Roberval balance 79, *79*, 149
   spring scales 79
wheel-lock muskets *44*, 45, *45*
Wickham, Henry 24
Willis, Thomas 105
Willoughby, Francis 90
Wint, Peter de *112*
Wittgenstein, Ludwig 57
Woolsthorpe Manor *98*, 99, 100
World Wars I and II 35, 60, *60*, 88

YZ
yeasts 74
zoology 93, 152
Zumpe, Johannes 122, *122*

Picture credits

Front cover: main image, Isaac Newton experimenting with light, Science Photo Library; inset: a reproduction of John Hadley's octant, Cosmos/Science Museum/SSPL. **Spine**: early variety of beet, AKG-Images. **Back cover**: a Roberval balance, Leemage/ Bianchetti.
Page 2, left to right, top row: Roger-Viollet/ J Boyer; Cosmos/Science Museum Library/SSPL; Bridgeman Art Library/Giraudon/RMN/The Trustees of the British Museum; 2nd row: Leemage/ Bianchetti; Dagli Orti/E Tweedy; Dagli Orti/Museo Nazionale di San Martino, Naples; 3rd row: Bibliothèque nationale de France/res livres rares 4-TE142-130; Corbis/Historical Archives; Cosmos/Science Museum Library/SSPL; bottom row: Leemage/Costa; Cosmos/Science Museum/SSPL; Cosmos/Science Museum Library/SSPL.
Pages 4/5: John Parker; 6r: Roger-Viollet/ J Boyer; 6tr: Leemage/Fototeca; 6c: AKG-Images/Ullstein Bild; 7t: Leemage/Bianchetti; 7c: Bibliothèque nationale de France, Cartes et plans, GE A282; 7br: Corbis/The Gallery Collection; 8l: Dagli Orti/National Palace, Mexico/G Dagli Orti; 8cr: Leemage/Aisa; 8tr: Leemage/Costa; 9l: Cosmos/Science Museum/SSPL; 9tl: Loomage/ Costa; 9br: RMN/Musée de l'Armée/J Y and N Dubois; 10t: Dagli Orti/The Art Archive; 10l: Leemage/PrismaArchivio; 10/11c: RMN/The Trustees of the British Museum, London; 11t: Bibliothèque nationale de France/res livres rares, res M-V-248; 11c: Cosmos/Science Museum Library/SSPL; 11r: Bridgeman Art Library; 12t: Leemage/Bianchetti; 12bl: Cosmos/Science Museum Library/SSPL; 12b: Cosmos/SPL; 12/13tr: Cosmos/Royal Astronomical Society/SPL; 13c: Dagli Orti/E Tweedy; 13tr: Bibliothèque nationale de France/res livres rares 4-TE142-130; 14tl & c: Bridgeman Art Library; 14tr: Cosmos/M Kulyk/SPL; 15tl: AKG-Images; 15r: Cosmos/Science Museum Library/SSPL; 15b: Leemage/Heritage Images; 16tl: AKG-Images; 16tr: AKG-Images/E Bohr; 16/17tr: RMN/Metropolitan Museum of Art/Image of the Museum; 17tl: Dagli Orti/Biblioteca Nacional, Lisbon/G Dagli Orti; 17tr: Cosmos/ Science Museum Library/SSPL; 17br: Leemage/ Photo Josse; 18/19: Leemage; 20l: Leemage/ Selva; 20r: Roger-Viollet/J Boyer; 21tl & c: Dagli Orti; 22tr: RMN/J G Berizzi; 22b: Leemage/Selva; 23t: Corbis/O Franken; 23b: Corbis; 24: Bridgeman Art Library/Archives Charmet; 25l: Jupiter Images (c) 2010; 25r: Cosmos/D Telemans; 25bl: Leemage/Farabola, 20tr Loomago/Fototeca; 26b: Dagli Orti/Bibliothèque des arts décoratifs, Paris/G Dagli Ort; 27c: AKG-Images/Nimatallah; 27br: Leemage/Costa; 30hd: Leemage/Heritage Images; 28bl: Bibliothèque nationale de France, Ms arabe 2509 fol, 40v-41; 28/29: Leemage/ Infatti; 29b: Leemage/ Bianchetti; 30bl: AKG-Images/L Lecat; 30/31tr: Cosmos/Science Museum Library/SSPL; 31r: Bibliothèque nationale de France, Cartes et plans, GE A282; 32tr: Dagli Orti/Musée Carnavalet, Paris/G Dagli Orti; 32bl: AKG-Images/Ullstein Bild; 33cl: AKG-Images; 33r: Leemage/Selva; 34tl: Dagli Orti/Science Museum, London/E Tweedy; 34b: Dagli Orti/Smithsonian Institute/E Tweedy; 35tl: Leemage/Costa; 35r: Corbis/EPA/D Ignacio; 36tl: AKG-Images; 36br: Corbis/Bettemann; 37: Leemage/Aisa; 38tl: Corbis/J Sugar; 38tr & tl: Cosmos/Science Museum/SSPL; 39br: Dagli Orti/National Palace, Mexico/G Dagli Orti; 40tl: Bibliothèque nationale de France/res 4-TB36-2; 40r: Dagli Orti/Library of the University of Istanbul/G Dagli Orti; 41tr: Cosmos/ Kulyk/SPL; 41/3bl: AKG-Images; 42t: Leemage/MP; 42b: Bridgeman Art Library/Archives Charmet; 43t: Corbis/The Gallery Collection; 43b: Leemage/

Costa; 44: RMN/Musée de l'Armée/E Cambier; 45t & c: RMN/Musée de l'Armée/J Y & N Dubois; 46tl & b: RMN/Musée de l'Armée/P Segrette; 46tr: RMN/G Blot; 47tr: Cosmos/Lambert/SPL; 47bl: Dagli Orti/Historical Museum of Vienna/ A Dagli Orti; 48c & b: Cosmos/Science Museum Library/SSPL; 49: Bridgeman Art Library/Archives Charmet; 50tl: AFP/HO/X-Tech Group; 50b: AFP/ L Gouliamaki; 51t: Cosmos/Science Museum Library/SSPL; 51c: DR; 52t: Leemage/Costa; 52b: CDDS/Bernand/P Gely; 53t: Cosmos/Science Museum, London/SPL; 53c: Cosmos/Science Museum/SSPL; 54 Leemage/Aisa; 55t: Cosmos/ Science Museum/SSPL; 55b: Cosmos/SPL; 56: Leemage/Photo Josse; 56/57tr: Leemage/ PrismaArchivio; 57c: Cosmos/J Burgess/SPL; 57tr: RMN/G Blot; 58: Cosmos/Science Museum Library/SSPL; 59tl: Corbis/D Lees; 59c & br: Leemage/Costa; 60tl: Cosmos/Science Museum Library/SSPL; 60b: Corbis/Hulton-Deutsch Collection; 61t: Corbis/K Hackenberg/Zefa; 61b: AKG-Images; 62tr: Dagli Orti; 62bl & c, 63: RMN/The Trustees of the British Museum, London; 64bl: Dagli Orti/Musée historique lorrain, Nancy/ G Dagli Orti; 64cr: Dagli Orti/Musée Carnavalet, Paris/G Dagli Orti; 65t: AKG-Images; 65br: Leemage/Photo Josse; 66tl: Jupiter Image (c) 2010; 66bl: Leemage/Bianchetti; 67: Cosmos/ Sciences Museum/SSPL; 68tl: Leemage/ Index/Pizzi; 68c: Leemage/Costa; 68br & 69tl: Bibliothèque nationale de France/res livres rares, res M-V-248; 69tr & 70bl: Cosmos/Science Museum/SSPL; 70c: Cosmos/K Guldbransen/SPL; 71tl: Cosmos/Science Museum/SSPL; 71b: AKG-Images/Ullstein Bild; 72: Cosmos/J Reader/SPL; 73t: Cosmos/Science Museum/SSPL; 73br: Bibliothèque nationale de France/R-6720 page 168; 74tl: Cosmos/Innerspace Imaging/SPL; 74tr: Bibliothèque nationale de France/R-6720 page 161; 74bl: Cosmos/SPL; 75: Bridgeman Art Library; 76bl & br: Bridgeman Art Library; 77tl: Bridgeman Art Library; 77tr: Dagli Orti/Musée Carnavalet, Paris/G Dagli Orti; 78bl: Roger-Viollet; 78t: Cosmos/A Hart-Davis/SPL; 79t: Leemage/ Bianchetti; 79br: Codemes/DR; 80tl: AKG-Images/ Nimatallah; 80bl: Scala Florence/HIP; 81tr & c: Cosmos/Science Museum/SSPL; 81b: Cosmos/ Ducros/SPL; 82: Dagli Orti/E Tweedy; 83tl: Cosmos/Science Museum Library/SSPL; 83tr: Leemage/Aisa; 83b: Cosmos/Royal Astronomical Society/SPL; 84tl: Cosmos/SPL; 84tr: Bridgeman Art Library/Courtesy of the Warden and Scholars of Oxford University; 85tl: Cosmos/SPL; 85br & 86tl: Cosmos/Science Museum/SSPL; 85tr: Leemage/Selva; 86bl: Corbis/R Ressmeyer; 87t: Cosmos/Nielsen/SPL; 87b: Cosmos/Science Museum Library/SSPL; 88tl: NLC Collection; 88br: Bridgeman Art Library; 89tl: Leemage/ L Ricciarini; 89br: Dagli Orti/Museo Nazionale di San Martino, Naples; 90tr: Bibliothèque nationale de France/est JC-8(A) pet fol, n° 1524 fol 64; 90bl: Cosmos/Science Museum Library/SSPL; 91tl: NLC Collection; 91c: Bridgeman Art Library; 91b: Bibliothèque nationale de France/res livres rares 4-TE142-130; 92tl: Dagli Orti/E Tweedy; 92br: Leemage/Aisa; 93t: Leemage/Costa; 93b: Leemage/Infatti; 94c: Cosmos/Royal Astronomical Society/SPL; 94bl: Leemage/Heritage Images; 95t: Cosmos/J Sanford/SPL; 95r: Cosmos/SPL; 95br: Bridgeman Art Library; 96c: Cosmos/ American Institute of Physics/SPL; 96br: Cosmos/NASA/ESA/STScI/SPL; 97: Bridgeman Art Library/Yale Center for British Art, gift P Mellon; 98tl: Cosmos/SPL; 98b: AKG-Images/E Lessing; 99b: Dagli Orti/E Tweedy; 99r: Leemage/Fototeca; 100tl: AKG-Images/E Lessing; 100b: Leemage/ Bianchetti; 101tl: Corbis/P C Chauncey; 101br: Corbis/Historical Archives; 102: Dagli Orti/British Library, London; 103t & cr: Cosmos/SPL; 103b: Dagli Orti/Biblioteca Marciana, Venice/A Dagli Orti;

104tl: Bridgeman Art Library; 104tr: Bibliothèque nationale de France/GR fol-TA9-253; 104b: Bridgeman Art Library/Archives Charmet; 105: Cosmos/Science Museum Library/SSPL; 105br: CDDS/Bernand/P Gely; 106tl: Cosmos/M Kulyk/ SPL; 106b: Musée Gustave- Flaubert d'histoire de la médecine, Rouen/J Petitcolas; 107t: Leemage/ J Bernard; 107b: Cosmos/E Kapitza/Focus; 108: AKG-Images; 109tl: Musée des Arts et Métiers, Paris; 109tr: AKG-Images; 109br, 110l, 110tr, 111t & br: Cosmos/Science Museum/SSPL; 112tl: AKG-Images/Electa; 112tr: Bridgeman Art Library/Yale Center for British Art, gift P Mellon; 112b: Leemage/Photo Josse; 113c: Bridgeman Art Library; 113b: Corbis/C Garrat/Milepost921/2; 114bl: Cosmos/Science Museum Library/SSPL; 114/115tr: Corbis/Ch O Rear; 115: Cosmos/SPL; 116tl: Dagli Orti/E Tweedy; 116bl: Cosmos/ Science Museum/SSPL; 117t: Leemage/Heritage Images; 117b: Corbis/R Hirshberger/DPA; 118tl: Cosmos/Science Museum Library/SSPL; 118tr. Ciel et Espace/Fujii; 119tr: AKG-Images/E Bohr; 119br: Corbis/Mosaic Images; 120c & bl: AKG-Images/Rabatti-Dominginie; 121tl: AKG-Images; 121tr: RMN/Metropolitan Museum of Art/Image of the Museum; 122t: Scala Florence/V&A/Images/ Victoria and Albert Museum, London, 122c: Dagli Orti; 122bl: RMN/J Lippe/BPK; 123tl: AKG-Images; 123br: Cosmos/Ch Irigang/Focus; 124bl: AKG-Images/E Lessing; 124tr: AKG-Images; 125r: Cosmos/A Lichtenstein/Aurora; 125b: Bridgeman Art Library; 126/127b: Leemage/Photo Josse; 126c: RMN/G Blot; 127: Bibliothèque nationale de France/med Louis XIV SR 669 revers; 128tl: AKG-Images/Cameraphoto; 128b: Leemage/Photo Josse; 129c: Leemage/Costa; 129cr: Ch Audebert/ E-Tuner; 129bl: Sipa Press/Collection Ribière; 130: F Zeri; 131tr: AFP; 131c: Jupiter Images (c) 2010; 131bl: Corbis/ER Production/Brand X; 131tl: RMN/BPK/DR/D Katz; 132tr: Bibliothèque de l'Institut, Paris; 133: Bibliothèque nationale de France/RES AA4 Le Blo; 134tl: Cosmos/Science Museum/SSPL; 134bl: Bridgeman Art Library/ Royal Naval Museum, Portsmouth; 135bl: Cosmos/S Percival/SPL; 135r: Bridgeman Art Library; 136tl: Museum of Art and History, Neuchâtel, Switzerland; 136b: AKG-Images/ Metropolis; 137tr: Leemage/Gusman; 137c: Cosmos/Science Museum Library/SSPL; 137bl: Cosmos/SPL; 138bl: Dagli Orti/Biblioteca Estense, Modena/A Dagli Ort ; 138tr: AKG-Images; 139cl: RMN/BPK/DR; 139tr: Leemage/Gusman; 139b: Dagli Orti/Biblioteca Nacional, Lisbon/ G Dagli Orti; 140bl: Leemage/Electa; 140/141tr: Leemage/ Photo Josse; 140c: Bridgeman Art Library/Yale Center for British Art, gift P Mollon; 141: Leemage/North Wind Pictures; 142tl & b: Bridgeman Art Library; 142cl: Bridgeman Art Library/Charmet; 143: Leemage/Aisa; 144/145b: Leemage/Photo Josse; 146tl: Dagli Orti; 146bl: Dagli Orti; 146r: Leemage/Aisa; 147t: Bridgeman Art Library/Archives Charmet; 147bl: Dagli Orti/Musée Carnavalet, Paris/G Dagli Orti; 147bl: AKG-Images; 148c: Cosmos/Science Museum/SSPL; 148b: Dagli Orti/Historical Museum of Vienna/A Dagli Orti; 148br: Leemage/Costa; 149tl: Leemage/Index/Pizzi; 149bl: Leemage/Frassineti/Agf; 149r: Roger-Viollet; 150bl: NLC Collection; 150c: Musée des Arts et Métiers, Paris/photo S Pelly; 150br: Bridgeman Art Library; 151bl: Leemage/Selva; 151c: Corbis/Historical Archives; 151br: Cosmos/Science Museum Library/SSPL; 152bl: Leemage/Aisa; 152br: Museum of Art and History, Neuchâtel, Switzerland; 153bl: AKG-Images; 153tl: Cosmos/Science Museum Library/SSPL; 153br: AKG-Images.

Illustrations on pages 44, 45, 69, 85 and 135 by Jean-Benoît Héron.

THE ADVENTURE OF DISCOVERIES AND INVENTIONS
The Age of Reason – 1600 to 1750
is published by The Reader's Digest Association Limited,
11 Westferry Circus, Canary Wharf, London E14 4HE

The book was translated and adapted from *L'Âge de Raison des Inventions*,
part of a series entitled L'ÉPOPÉE DES DÉCOUVERTES ET DES INVENTIONS,
created in France by BOOKMAKER and first published by Sélection du
Reader's Digest, Paris, in 2010.

Translated from French by Tony Allan

Series editor Christine Noble
Art editor Julie Bennett
Designer Martin Bennett
Consultant Ruth Binney
Proofreader Ron Pankhurst
Indexer Marie Lorimer

Colour origination Colour Systems Ltd, London
Printed and bound in China

READER'S DIGEST GENERAL BOOKS
Editorial director Julian Browne
Art director Anne-Marie Bulat
Managing editor Nina Hathway
Head of book development Sarah Bloxham
Picture resource manager Sarah Stewart-Richardson
Pre-press account manager Dean Russell
Product production manager Claudette Bramble
Production controller Sandra Fuller

We are committed to both the quality of our products and the service we provide to our
customers. We value your comments, so please feel free to contact us on 08705 113366
or via our website at **www.readersdigest.co.uk**

If you have any comments or suggestions about the content of our books, you can
email us at **gbeditorial@readersdigest.co.uk**

CONCEPT CODE: FR0104/IC/S
BOOK CODE: 642-004 UP0000-1
ISBN: 978-0-276-44516-3
ORACLE CODE: 356400004H.00.24